Animal Viruses and Humans

Animal Viruses and Humans

A NARROW DIVIDE

How Lethal Zoonotic Viruses
Spill Over and Threaten Us

WARREN A. ANDIMAN

PAUL DRY BOOKS
Philadelphia 2018

First Paul Dry Books Edition, 2018

Paul Dry Books, Inc.
Philadelphia, Pennsylvania
www.pauldrybooks.com

Printed in the United States of America

ISBN 978-1-58988-122-8

To my wife, Marie,
and my daughters, Sarah and Alexis,
whom I love and greatly admire
for their intelligence and selfless devotion
to their fellow human beings.
They have always encouraged me
to do my best work
and taught me to seek after
my better self.

CONTENTS

Everybody knows that pestilences have a way of recurring in the world; yet somehow we find it hard to believe in ones that crash down on our heads from a blue sky.

Albert Camus, *The Plague*

That human health and Earth's health are intertwined can sound like a truism—more of a bumper sticker or poetic truth than scientific fact. Yet a growing body of research suggests that disrupted ecologies may indeed produce more disease.

Brandon Keim, *Anthropocene Magazine*,
December 14, 2016

All is flux, nothing stays still.

Heraclitus, c. 540–480 B.C.

PREFACE

Viruses are obligate intracellular parasites; that is, they can replicate only within the confines of living animal or plant cells, bacteria included. Once they gain entry to a cell's interior, they hijack many of its rich biochemical offerings and spread by one of three ways: by producing hundreds or thousands of progeny that bud from the cell's surface membrane; or by inducing the fusion of adjoining cells so they can move into neighboring property as the membranous barrier between cells melts away. Some viruses completely destroy the cells in which they reproduce and thereby find sanctuary in the juices between cells or in the blood and lymph, whereupon they can move to wherever the body's life blood (literally) carries them.

Thousands of known virus species have evolved since life first appeared on Earth. They have adapted their parasitic lifestyles to thousands of living species. Their sole purpose is to reproduce and spread, as is the case for many higher life forms. As a result, some have developed the capacity to "jump" from one species to another, thereby expanding their ecological range in true Darwinian fashion as they continue to mutate and slowly acclimate to new intracellular milieux.

The arboviruses have found a way to travel among animal species by hitching a ride inside insects. These highly mobile vectors of disease take a blood meal from their animal reservoirs—cattle, pigs, birds, bats, monkeys, etc.—and transmit teeming blood-borne virus herds to other animals. Some of these animals happen to be humans who find themselves in proximity to the creatures that passively serve as the source of their infection. In contrast, some viruses travel directly from animal to animal or from animal to human being without an insect interloper. (A famous story tells of Joseph Meister, the boy who acquired rabies from the bite of a dog and who was cured by Louis Pasteur in 1885; no mosquitoes, ticks, or mites played a role.) The journey among animals is eased when people, sub-human mammals, and birds live or work in close quarters. Because vocations, recreations, and ecological niches vary greatly from place to place, the kinds of animal vectors and the conditions under which virus transmission occurs also vary greatly.

The viruses and diseases, the zoonoses, I've chosen to write about in this book are characterized by a direct mode of the spread of infection from animals to humans, and by the great variety in place of origin and the mix of animal species involved. These examples are meant to describe only a small sample of what is already out there but, more menacingly, what is inevitably on its way, in forms we can only imagine.

HIV, the human immunodeficiency virus, is one of the most recent and, doubtless, the most terrifying of these zoonoses. Several readers of early drafts of this book asked why I did not include HIV. My response, perhaps wrong-headed, was that HIV/AIDS and its history of pestilence and death were already familiar to many people and that I wanted to expand their horizons by providing stories of other zoonoses—ones that have already maimed and killed and that might someday break free, should the right combination of brief encounters take place.

HIV-induced disease is a paradigm of a series of random

animal-to-human transmission events. There are actually two AIDS viruses, HIV-1 and HIV-2, both members of the lentivirus family, a group of pathogens shared by monkeys, apes, and human beings. It's believed that a sequence of significant cross-species lentivirus transmissions began only in the past century or two, this despite the fact that Old World simian lentiviruses have evolved along with their host species for millions of years. There are at least ten simian immunodeficiency viruses, each related to a different simian species. SIVcpz, the chimpanzee virus, and SIVgor, the gorilla virus, are the two candidates that ultimately made their way to humans. Transmission is thought to have occurred through cutaneous or mucous membrane exposure to ape blood or other body fluids in the course of hunting and eating bushmeat.

A number of worthy writers and journalists have covered aspects of these sorts of events in the last several decades. Much of their focus has been on the social and environmental conditions that create the settings in which viruses can easily jump from animals to humans and then, more ominously, among their human hosts. These writers have sometimes targeted viruses whose very names are so familiar as to make us stand at attention: Ebola, avian influenza, and Lassa fever, among others. But as a physician trained in clinical and diagnostic virology and as a clinician on the front lines, I felt obliged to introduce readers to the magical and mysterious world of the viruses themselves: how they look and how their very structure and molecular make-up allow them to parasitize the cells of their mammalian and avian hosts so efficiently—and thereafter, to cause more generalized disease. Having invaded their prey, how do viruses then take hold of the cells' complex machinery in order to fulfill their sole raison d'être, that is, to reproduce promiscuously and to wreak havoc wherever they can find a home? In addition, I describe some of the physical manifestations of viral attack. I do this with some trepidation because the pictures I paint are far from pretty. Many of us get a bit of a kick out of being scared and even horrified, so I believe there is some value in having readers understand the consequences of allowing these diseases to cross international borders and to move uninhibited into our hospitals, neighborhoods, homes, and bodies. The recent epidemic of Ebola virus disease has given us a glimpse into the terrors unleashed by these super-sub-microscopic beasties. Finally, I will briefly revisit some of the lifestyles and habits that bring members of our own species into intimate contact with representatives of other species. We not only depend on animals for food, which brings us close to their very flesh and blood, but we also love animals, which brings us into contact with their secretions and excretions in a manner not unlike that of the humans we love and nurture.

As I tell my stories I also take the liberty of conveying some basic principles of virus disease epidemiology and pathogenesis. Many of these principles relate to the majority of infectious agents, including bacteria and fungi, but I have placed them in the text when elements of the story provided an opportunity for a brief encounter with one of these overarching principles. There are five basic tenets that explain the behavior of infectious diseases: i) contact between a microbe and its host does not necessarily result in infection; ii) the earliest stages of infection do not necessarily result in clinically apparent illness (that is, some infections are subclinical); iii) the manifestations of clinical illness can range from very mild to severe; iv) the clinical expression of disease depends, to a varying extent, on the host's immune response to the pathogen; and v) the pathogen and the host share a geographic locale and an ecological niche, both of which play a role in determining the host's susceptibility and the extent of clinical risk.

Finally, I have chosen to assume an informal style in my writing. I am a physician-scientist who in my daily discourse uses all kinds of polysyllabic words and arcane phraseology. I have labored to adapt my style to one that can easily be approached by educated readers, who can, if they wish, supplement their reading of these chapters with further etymological and scientific searches. I have made every attempt in these transformations of language not to distort scientific fact. If by chance I have transgressed in these attempts, I ask my colleagues and more specialized readers to forgive me.

I also ask to be forgiven for my tendency to anthropomorphize the viruses' appearance and behavior. I worked with these miniature beasties in the laboratory for thirty-five years, and I've cared for patients infected with them for so long that I have begun to regard them as having motives (mostly nefarious) and bodies (mostly startlingly beautiful) that seem almost humanoid, or perhaps even sentient and highly motivated.

CHAPTER 1

Millions of Tiny Crowns Wreak Havoc in Two Kingdoms (Or Never Wipe a Camel's Nose)

MERS (MIDDLE EAST RESPIRATORY SYNDROME)

> CDC is working closely with the World Health Organization (WHO) and other partners to better understand the public health risk presented by a recently detected, novel coronavirus. This virus has been identified in two patients, both previously healthy adults who suffered severe respiratory illness. The first patient, a man aged 60 years from Saudi Arabia, was hospitalized in June 2012 and died; the second patient, a man aged 49 years from Qatar with onset of symptoms in September 2012 was transported to the United Kingdom for intensive care. He remains hospitalized on life support with both pulmonary and renal failure.
>
> *Morbidity and Mortality Weekly Report*, Vol. 61, No. 40, October 12, 2012

In June 2012, a 60-year-old Saudi man was admitted to a hospital in Jeddah. He complained of week-long symptoms of productive cough, fever, and shortness of breath. He had no history of chronic diseases affecting his lungs or kidneys. Examination of his chest X-ray revealed opacities in both lungs, inflammatory fluids in the tissues surrounding them, and diminished lung volume. These abnormalities progressed over the next several days despite treatment with a cocktail of anti-bacterial, antiviral, and anti-fungal antibiotics prescribed for him hurriedly even in the absence of a definitive diagnosis. (There was no test for the agent that infected him.) Initially, three different pathogenic bacteria were isolated from his lung secretions, but none was believed to be causing his illness. The patient required mechanical ventilation to keep his tissue oxygen levels normal. Days later, quite unexpectedly, his kidneys began to fail; on his eleventh day in the hospital, he died of progressive pulmonary and renal failure.

Analysis of the man's sputum by PCR (polymerase chain reaction, the genetic technique commonly used to analyze and identify the source of blood at a crime scene, or of semen in a rape victim) revealed that he had been infected with a coronavirus. Another test proved that the man had antibody to the virus in his blood, but clearly not enough to quell advancing disease. To everyone's surprise, blood samples from 2,400 other Saudis ("controls") were found not to contain antibodies to the new virus, indicating that the poor dead man had been infected with a virus that very few, if any, of his countrymen had ever been exposed to.

After sufficient quantities of this coronavirus had been grown in cell culture, exhaustive genetic analysis of the viral genome was undertaken. It was discovered that the new virus belonged to a lineage called beta coronaviruses and, oddly, that its closest relatives were coronaviruses that commonly persist silently in bats. Stranger still, the new virus was more closely related to bat coronaviruses than to the other coronaviruses already known to infect humans. This finding suggested strongly that one or more other animal species, certainly including bats, were the likely reservoir.

Corona is a Latin word meaning "crown," from the an-

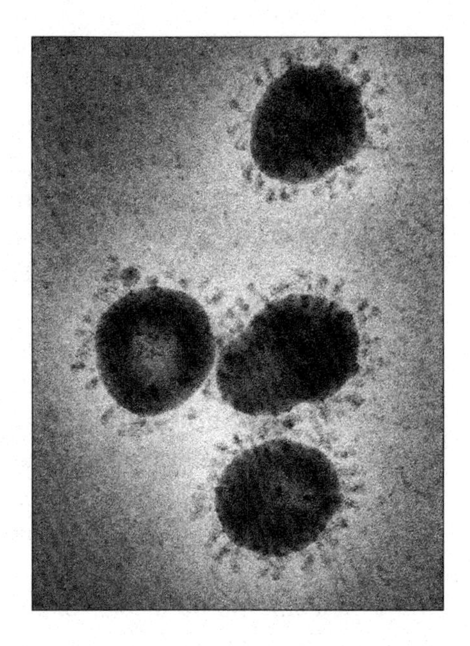

Transmission electron micrograph of coronavirus particles. Note the crown-like (corona) or halo-like appearance. Coronaviruses are responsible for SARS, the severe acute respiratory syndrome, and MERS, the Middle East respiratory syndrome. Content provider Dr. Fred Murphy of the Centers for Disease Control and Prevention and the Public Health Image Library (PHIL).

cient Greek, korōnē ("garland, wreath"). Each coronavirus particle is surrounded by a garland of club-shaped projections, creating the appearance of a daisy or a crown. The projections comprise sugar and protein moieties which together form the "keys" that latch onto receptors on the host cell's outer membrane. This lock and key interaction determines the tissue specificity of the virus. For example, the cells that bear these coronavirus receptors are found in the mucous membranes of the nose and trachea, as well as the kidney and some fibrous tissues. Based on this very specific virus-tissue interaction, it is not surprising that the Saudi man's respiratory and urinary tracts were the systems most dramatically affected and the first to fail.

A year and a half after the first unfortunate Saudi man died, a forty-three-year-old countryman of his, also from the Jeddah region, went to the hospital, complaining of shortness of breath. This followed a week-long illness characterized by runny nose, cough, and malaise. The patient owned a herd of nine camels. The camels were housed about forty-five miles south of Jeddah, where he and a group of friends visited the animals every day. Coincidentally, the men noticed that four of the camels had runny noses and that one was being treated by their sick friend, who had been gently applying a medicinal salve to the camel's nostrils. This kind man developed irremediable lung disease and died two weeks after being admitted to the hospital.

The patient's eighteen-year-old daughter developed an upper respiratory infection some time during her father's hospital stay, but she recovered fully after three days. None of the dead man's friends got sick. There's an important principle at work here—inoculum size: that is, the density of pathogenic virus particles in a particular body fluid which later find a welcome home in one or more of the body's portals of entry—for example, nose, mouth, throat, urethra, rectum, damaged skin, or conjunctiva. The dead man who tended to his sick camel's URI by direct contact with the animal's nasal discharge was clearly put at greater risk than his buddies by having handled the camel's mucus, a slimy inoculum later found to be teeming with virus. His friends merely visited the camels, and his daughter only lived in the same house with him. One can easily imagine that a simple, unconscious movement of the dead man's mucus-contaminated finger from the camel's nose to his own transferred as many as thousands or millions of virus particles from one juicy home to another. Remember, viruses are very, very, very small. (One lusty sneeze laden with influenza or common cold viruses can easily contaminate half an auditorium or most of an airplane cabin, as thousands of viruses seek a welcoming human landing platform.)

The disease caused by this camel (or bat) coronavirus was given the name Middle East Respiratory Syndrome, or

MERS. (It bore some resemblance to an earlier disease with a similar acronymic name, Severe Acute Respiratory Syndrome, or SARS, one with a more alarming and widespread influence. See chapter 5.) Exhaustive genomic analysis of nasal fluid collected from the drippy-nosed camel and from the patient on four different hospital days showed that the RNA of all five samples were identical. In contrast, no virus was found in the nose of the dead man's daughter and none was found in the milk, urine, or feces of any of the camels. In this particular illness cluster, the MERS virus, like many other coronaviruses known to us, appeared to share a proclivity for infecting cells of the respiratory tract. When the structure of the camel coronavirus RNA was compared to dozens of other coronavirus RNAs in the GenBank database, it was found to be closely related to coronavirus isolates from Riyadh, Munich, Abu Dhabi, England, France, Jordan, the United Arab Emirates, and Qatar. Coronaviruses in this branch of the coronavirus family tree were already widely dispersed in parts of Europe and the Middle East, yet no serious outbreaks occurred. It was a mere handful of fatal respiratory illnesses in these Saudi men, a few others, and several camels, that brought MERS to the world's attention. During the following year, about eighty cases of MERS were identified; a little more than half of those afflicted died, for a case fatality rate of fifty-six percent. Based on extensive understanding of the MERS coronavirus, that is, MERS-CoV, and its molecular structure and behavior in vitro, we now know that it spreads from animals to humans, but also among humans in very close contact with one another, within households, for example, or in hospital settings, as with other respiratory and gastrointestinal viruses. Most of the patients identified have had underlying diseases such as diabetes or end-stage kidney disease, suggesting that the MERS coronavirus takes advantage of people with some degree of immune compromise.

And some minority of the patients appear to have been "super shedders," each one secreting and then transmitting the infection to more than a half-dozen of their close contacts.

The early months of 2014 were characterized by a continuous ebb in the number of confirmed MERS cases that came to light. No more than twenty or thirty samples of respiratory secretions were received each day by the Jeddah Regional Laboratory. The Jeddah Laboratory is a reference facility governed by the Saudi Ministry of Health and does preliminary and confirmatory testing for MERS-CoV for laboratories throughout the kingdom of Saudi Arabia. Infections consistent with MERS-CoV were springing up all over, but most were focused on Jeddah. The third week of March 2014 saw an abrupt increase in samples submitted, which continued to escalate through most of April. The reference lab began receiving somewhere between 200 and 400 respiratory specimens a day, and, as before, the majority came from patients living in and around Jeddah. As the outbreak advanced, the lab served not only as a diagnostic facility but also as the hub of a detailed epidemiologic investigation for which the submitted specimens served as source material. The specimens allowed investigators to follow webs of transmission and, perhaps, would yield some clues about the virus's virulence. The scientists also believed they might discover new sources or a greater frequency of animal to human, that is, zoonotic transmissions, and maybe even instances of nosocomial transmission, that is, spread from person to person within the hospital setting. Of the more than 6,200 samples that were tested in Jeddah between January and the end of April 2014, ninety-three percent were received after March 26. But of that number, only 168 samples, or three percent, contained MERS-CoV. This may appear to be a shockingly small number, but with diseases like MERS, which has no unique clinical presentation (coughs, colds, pneumonia, and fevers are ubiquitous), and which is bound to cause exaggerated fretfulness in specific populations at risk, doctors are obliged to submit many more specimens for analysis than are expected to be positive. After all, as healthcare workers, we do not want to miss any bona fide infections, especially potentially fatal ones, which are most likely to respond to intervention in their early stages.

Half the positive samples came from King Fahd Hospital, a clear signal of nosocomial spread. KFH serves many of the expatriates who flock to the city to find employment. Quite logically, there were more samples taken from contacts of patients identified in Jeddah, the outbreak's epicenter, than from contacts traced to other parts of the country. Not surprisingly, the secretions from asymptomatic close contacts of patients carried lower titers of virus than those from symptomatic patients themselves, because persons who find their way to doctors' offices or hospitals are more likely to be symptomatic and symptomatic patients are more likely to be carrying greater amounts of virus. Viral load usually correlates with degree of illness and a greater likelihood of infectiousness, and hence, transmissibility.

The genomes of a representative number of virus isolates from the Jeddah outbreak were fully sequenced. That is, every genetic signature in the chain of sequences that constitute the complete genetic code of each virus isolate was identified. They showed that all the Jeddah viruses clustered in a single small genetic family, or clade. In contrast, the viruses from Riyadh were a bit more genetically diverse. One clade from the Riyadh family of viruses showed internal similarities that suggested person-to-person transmission. One person carried the Riyadh MERS virus back to Indiana, in the United States, and a few other carriers exported the virus to the Netherlands. Only one coronavirus isolate collected in Riyadh was similar to those that clustered in Jeddah. The patient who carried this viral isolate had visited his hospitalized son in King Fahd Hospital in Jeddah before returning to his home in Riyadh, but there were no reports of further family household spread.

When subjected to the most detailed forms of genetic analysis, MERS-CoV isolates collected from camels in Taif, Jeddah, and Qatar were found to resemble closely human isolates from Riyadh and Bisha. Taif is southeast of Jeddah and Qatar is on the eastern edge of the Arabian Peninsula. I present these seemingly obscure geographic details to illustrate how efficiently respiratory and gastrointestinal viruses cross both domestic and international borders. A virus that likely leaked from the nose of a camel in Saudi Arabia traveled, with minor variations, through that country and soon found its way to the American Midwest and the Netherlands and Greece. We now know that MERS-CoV was widely distributed in camels throughout Arabia at least since late 2013.

Luckily, although small clusters of MERS cases occurred in both hemispheres, the disease never spread widely; there has been no epidemic. Isolated bits of data suggest that the titers of virus in respiratory secretions did not rise significantly in the course of human-to-human transmission, as sometimes happens when viruses slowly adapt as they pass from host to host. Also, unlike influenza virus, the MERS-CoV virus outer coat did not change very much as it moved from person to person, lending credence to the idea that specific antibodies raised in the course of one infection might also protect the host against subsequent infection by a closely related virus, increasing the levels of "herd immunity" in the general population.

The broader coronavirus family of viruses infects many animal species. Among the targets are humans, dogs, cats, cattle, rabbits, turkeys, and chickens, in which each coronavirus affects, primarily, parts of the respiratory tract. Pigs are pummeled in the GI tract and central nervous system, mice in the liver, and rats in the tear ducts and lymph glands of the head and neck. Coronavirus particles have been seen by electron microscopy in human feces, and some of these strains are related to calf diarrhea virus, providing compelling evidence of likely spread between ungulates and people. This interspecies spread is further supported by the fact that the first human coronavirus was isolated from a schoolboy suffering from a URI in 1965. The morphology of this early isolate closely resembled the avian infectious bronchitis viruses. Despite compelling evidence of animal-to-human transmission, it is not clear whether the converse holds true: can human coronaviruses spread to and infect animals of other species?

Coronaviruses spread disease throughout the world, and we believe that the great majority of healthy adults have at some time survived infection, beginning in early childhood. Nevertheless, coronavirus infections are less common than those caused by rhinoviruses (the "common cold virus") and influenza viruses, but they always skulk somewhere among the marauders that can be found in the throats, windpipes, and lungs of sneezing, coughing humans during the winter and spring. Also, the clinical manifestations produced by these multiple respiratory pathogens are indistinguishable. As with influenza and rhinoviruses, coronaviruses are also known to be responsible for some acute exacerbations in patients with asthma and COPD. When coronaviruses infect cells of the tracheal mucosa, degenerative changes are observed in the cilia, those gentle, waving, feathery cellular appendages that clear the mucus from our noses and airways.

Camels and humans have had an intimate and historically critical relationship going back at least fifteen centuries. Arab Bedouins have shaped their armies and nourished their people, supported by the virtues of their camel herds. The rise of the Islamic empires of the Middle East could not have succeeded without the foundations built upon the camel's back. So it's not at all surprising that some present-day Arabs continue to revere and love their camels to such a degree that they are apt to wipe a camel's nose and apply soothing salves much as any good parent does for a child with a sore, runny nose.

With concern growing about the probable spread of MERS throughout the Arabian Peninsula and beyond, the World Health Organization updated its guidelines and cautioned locals and international visitors to the kingdom to practice general hygienic measures, which included the recommendation to wash hands regularly before and after touching animals and particularly to avoid all contact with sick animals. Travelers were also warned to avoid consumption of raw and undercooked animal products. To provide greater specificity, WHO reminded persons at high risk for MERS—those with diabetes, kidney disease, and chronic lung disease—to avoid all contact with camels, to refrain from eating camel meat, and most especially, to eschew drinking raw camel milk and raw camel urine. It was advised that camel milk be boiled before drinking.

But why camel urine? For more than a thousand years, the health and beauty benefits of camel urine have been applauded in many Middle Eastern countries. In some Islamic nations various facial creams, hair oils, cosmetics, and aphrodisiacs have included camel urine as an essential ingredient. On occasion, it's also been promoted as a nostrum for certain cancer patients. Its proponents claim that camel urine is sterile and free of toxins and can be stored at room temperature for up to two weeks "without being corrupted."

According to some religious scholars, the health benefits of camel urine can be traced back to a verse in the Hadith, a Muslim holy book. The legend describes the healing benefits that camel urine afforded to members of a certain tribe who were ill-suited to the climate in Medina. On the advice of a prophet, they drank camel urine and became healthy.

The WHO admonition against contact with camel urine in any form has now been echoed by the Saudi agricultural ministry, a precedent in a country where camels are integral to the culture and are both lauded and revered.

Despite the MERS virus's relatively low transmission rate, two family clusters of MERS infection have been meticulously recorded in the medical literature, one occurred in Riyadh and the other in London. The English cluster originated in a UK resident who had recently returned from an eight-day sojourn in Saudi Arabia, during which time he made pilgrimage to Mecca and Medina. The traveler reported no known contacts with persons afflicted by serious respiratory illnesses, but within days of returning to the UK, the man's condition deteriorated rapidly and he received every conceivable form of intensive care, including the most up-to-date modalities of respiratory support. An initial viral culture revealed the presence of influenza A, but another, collected a week later, contained the MERS coronavirus. Among twenty household contacts of the index patient (a term used to identify the first person in a cluster shown to be definitely infected), six developed respiratory infections, but the MERS virus was isolated from only two. The two had also visited the index patient in the hospital, and like the index case, both had also been infected with a second respiratory virus, parainfluenza. One of the two had cancer and was receiving chemotherapeutic agents likely to have damaged his immune system. Despite heroic efforts, the traveler's illness progressed and he died on the eleventh hospital day. One of the MERS carriers, a woman who had direct contact with the hospitalized index case for a total of two and a half hours, survived, following nine days of a flu-like illness. The cancer patient also survived.

Intensive public health efforts were launched to identify and study all the close contacts of the dead man. Six persons among his seventeen household and healthcare contacts and hospital visitors developed colds and coughs, but none had MERS virus in respiratory samples. Similarly, among the twenty-five close contacts of the woman with the "flu," three developed self-limited respiratory illnesses, and none was found to have the MERS virus in samples taken from their respiratory tracts. Clearly, this particular virus strain in its original incarnation was not easily transmitted from person to person and did not acquire that particular characteristic in its subsequent travels. The degree to which a virus is transmitted person-to-person is the characteristic that public health officials look for, prior to deciding whether or not to press the panic button.

Some additional interesting observations emerged from the extensive investigation of another Saudi family cluster. Of the four MERS-infected family members, ages sixteen to seventy, two died: the index patient, an elderly man, who had diabetes, pulmonary edema, hypertension, and coronary artery disease, and a thirty-nine-year-old factory worker, the old man's son, who was a long-time smoker with a history of asthma. The four infected patients lived in an extended family compound comprising twenty-eight individuals; none of the two dozen others became ill. Almost all the males in the family ate together, but separate from their female relatives and children. Sleeping quarters were a hodge-podge of married men and their wives and children. There were no household pets or domestic animals, but one of the four MERS patients had attended the slaughter of a camel about ten days before he came down with fevers, shaking chills, cough, and night sweats. He recovered eight days after admission to the hospital for treatment of pneumonia. One MERS-infected woman attended to her husband's needs at home during the earliest days of his illness, but following hospitalization, she visited her husband only very infrequently. None of the nine children in the household became ill; as custom demanded, they did not tend to their fathers' medical needs, and they did not visit the hospital. None of the doctors and nurses who cared for the hospitalized men became ill. Some of the MERS patients had gastrointestinal symptoms, others did not. Two developed signs of renal failure; the others did not. All four developed significant drops in their blood cell counts.

These family case histories have refreshed our familiarity with a basic principle of infectious diseases: the same microorganism can produce a spectrum of clinical expressions, and conversely, particular infectious diseases can each be produced by an array of different bacteria and viruses. URIs, pneumonia, and kidney infections can each be caused by a litany of microorganisms. Also, individual humans vary widely in their susceptibility to infection by pathogenic viruses. These differences are largely the consequence of the unique composition of each of our immune systems and the special environments in which we live. Our immunity to infection is a reflection of our unique genetic make-up and our past history of exposure to a multiplicity of infectious agents, some naturally-acquired and some gifted to us in the form of attenuated vaccines.

The MERS coronavirus, MERS-CoV, a member of a family best known for causing respiratory diseases in a number of different animal species, can, under certain conditions, cause clinical pathology in the kidneys, the GI tract, and the blood. This phenomenon points to the fact that some viral species carry molecular arrays on their surfaces that interact with receptors borne on the surfaces of a variety of cell tissue types. Most of these receptors are there to serve entirely different biologic processes, but through eons of mutation and structural change, the viruses have accidentally—but luckily for them—acquired the keys that latch onto receptors carried by the cells they ultimately parasitize. As we observed, most of the MERS patients started out with mere colds or coughs or what we often refer to as flu-like illnesses, by which we mean total body symptoms such as fever, chills, joint and body aches, including headache and backache, and, most prominently, hacking cough. And some patients, as we have seen, were initially laid low by chronic diseases or other respiratory pathogens. In these cases, the MERS virus must have selfishly taken advantage of the hosts' weakened state, damaged tissue, and perhaps their altered immune status.

In these unfortunate few, the "flu" merely heralds a progressive disease that affects the lungs in a most serious way, necessitating the use of mechanical ventilators and even more. It's never been entirely clear why such a broad range of clinical scenarios resulting from infection by a single microorganism occurs. However, there are theories that are logical and that simply need more analysis. Perhaps these individual differences lie in our genes; our unique individual make-ups affect the ways our immune systems react to pathogen invasion. Some of us may mount very aggressive antibody and cell-mediated immune responses accompanied by an explosion of inflammatory mediators. Some of the antibodies that are meant to protect us may instead hurt us, for example, by joining with elements of the invaders, producing molecular complexes that can clog tiny blood vessels or become lodged in the interstices between cells. Some antibodies enhance rather than dampen the infection by causing the virus to be ingested by cells of the immune system, in which they grow exuberantly. Certain individuals may harbor variations in the structure of cell receptors to which the virus initially attaches; this may result in a more languid attachment rather than a robust one. Finally, it's easy to imagine that some victims of viral attack are exposed to much larger numbers of infectious particles than others. Compare the vulnerability of the subway train passenger standing just a few feet away from the juicy cough of a tuberculosis "supershedder," as compared to a fellow passenger at the other end of the subway car—or the tourist on a cruise ship who happens to touch the bathroom doorknob soon after a fellow passenger infected with norovirus touches the same doorknob, having neglected to wash his hands after using the commode. The viral load to which one is exposed is perhaps the most critical variable that determines who gets infected and who does not, or who gets a really nasty infection versus a mild one, or, finally, one who gets infected but remains totally asymptomatic. (This last situation is very common!) The person who wipes the nose of an acutely MERS-infected camel is much more likely to get sick, indeed very sick to the point of death, compared with a person who passes the camel's enclosure with a simple greeting of "salaam aleikum."

By February 2015, laboratory-confirmed cases of MERS had been identified in nine countries in the Middle East. Travel-associated cases were reported throughout Europe, along the entire coast of North Africa, and in Malaysia and the Philippines. There were 50 cases in Saudi Arabia in February 2015 alone. In toto, since 2012, there have been 1,126 confirmed cases reported by WHO, with 376 deaths. Eighty-five percent of all cases have been reported from Saudi Arabia. Only a handful of infections occurred among people visiting healthcare settings.

The belief that clinics and hospitals were places relatively free of patient-to-patient contagion, based on these numbers, was short-lived. In June 2015, reports from South Korea sparked alarm throughout the Far East. Once again MERS broke through its original regional constraints. The South Korean index patient ran a farm equipment company in Bahrain and returned from his periodic business trip on May 4.

Eight days later he came down with a cough and fever, but visited four health facilities without a definitive diagnosis before he got so sick that he was hospitalized in Pyeongtaek, a city forty miles southwest of Seoul. He was placed in a two-bed room, but he was not confined to his room. It was reported that he was often found sneezing and coughing in the hall. It was not surprising to learn that nurses and doctors in a Korean hospital had no knowledge of MERS and initially were not told of the man's trip to the Middle East. The sick businessman was transferred to another hospital where, during the admission interview, he reported his trip to Bahrain, which is not among the countries considered to be a hotspot for MERS. He failed to mention his side visits to Saudi Arabia and the United Arab Emirates, the actual crucible of the MERS epidemic, where over four hundred people had already died of MERS.

A number of secondary cases were traced directly to this man: his roommate during the original hospitalization and the roommate's son, who had dutifully visited his father, succumbed. The son was quarantined but violated his confinement and traveled to Hong Kong and mainland China. He got sick, was diagnosed with MERS, and was hospitalized in China. In the course of the exhaustive public health investigations in Korea it was discovered that still other people contracted MERS while visiting the same health facilities that the index patient had visited in his initial fruitless quest for a diagnosis and treatment. The patient's wife also contracted MERS, but her illness was mild and self-limited. After all was said and done, the health authorities concluded that the majority of cases were acquired only in the various health centers the patient had visited, not in the wider communities.

It's tough to know where to place blame for this unfortunate international outbreak. It's not surprising that persons working at a regional hospital in South Korea should be ignorant of MERS. As far as we know, there had been no previous cases of MERS reported in the Far East. People come to outpatient clinics all the time with fevers and cough. These symptoms are somewhere near the top of the list of complaints that bring patients to the doctor, and during flu season they are at the very top. Why would any such routine visit elicit a question about foreign travel, especially travel to the Middle East? Only if you'd heard of MERS would you be impelled to be that specific in your questioning. But in all fairness, it was probably unwise to place a febrile, coughing patient with a roommate and to allow him to roam the halls of the hospital. We don't know whether visitors or hospital staff were required to wear masks and gloves when caring for the patient. That would have been a reasonable precaution. It's completely unforgivable that a man on quarantine was able to break away from his restricted confines. In doing so, he boarded a plane and traveled to another country. We still don't know if anyone else on the plane or in Hong Kong contracted MERS, or if they did, whether they remained asymptomatic or got sick. In general, only a minority of people ill enough to be hospitalized will come to the attention of public health authorities, and then only if their illnesses appear on a list of "notifiable" infections, such as influenza during a flu epidemic or MERS during a period of international spread. I would venture to guess that most hospitals in China at the time were unable to diagnose MERS, and that most patients hospitalized with fever and a bad cough were treated with an antibiotic and given supportive care, maybe a little oxygen, maybe a little mist by mask, maybe some IV fluids. The discharge diagnosis in such cases would probably say, "lower respiratory tract infection, possibly viral, of unknown etiology."

If a patient like our South Korean businessman came to an academic institution, such as the one in which I work, I or one of my colleagues or trainees might be asked to do a consult if the patient was sick enough to cause more than routine worry, and if there was an urgent need to identify a specific etiology and specific treatment. It's part of our "scripted" rote questioning that in such cases we ask about foreign travel or travel to a distant region within the patient's native country. We ask about pets and any other recent animal exposures; we ask about recent ingestion of any unusual foods or atypical water sources; and we ask the patient if he's been exposed to anyone else who's sick. The answer to one of these questions sometimes provides a vital cue to a possible diagnosis, at least one worthy of thorough investigation through blood tests, radiographic procedures, or rarely, a more invasive procedure such as biopsy or endoscopy. It all depends on thorough detective work guided by a large, readily accessible fund of knowledge.

If MERS were to cause a serious outbreak in one of the industrialized countries of the world, we have sensitive and specific tests for the MERS-CoV antigens and antibodies, which are stocked at many American state health departments and at the Centers for Disease Control and Prevention. In contrast, it's unlikely that the poor countries of the world will have the funds and the laboratory capabilities to do surveillance and rapid diagnostic testing should MERS spread beyond those countries already affected.

CHAPTER 2

It's a Good Thing Pigs Don't Fly

SWINE INFLUENZA

> As of June 15, nearly 36,000 people in 76 countries have been infected with influenza A (H1N1) and 163 have died. Last week, WHO raised the pandemic alert level from phase 5 to 6—the highest level—officially signifying the start of the first influenza pandemic since 1968.
>
> *The Lancet*, Volume 373, June 20, 2009

> The Health and Human Services subcommittee held an emergency panel hearing on the swine flu at Capitol Hill. The panel agreed that currently it is not necessary to test people coming off planes or crossing the border from Mexico, where the outbreak has been most severe. Their rationale was that it takes 24 hours to five days before a patient shows signs of the swine flu, which could make individual testing of little use.
>
> *The Epoch Times, New England Edition*, April 30–May 6, 2009

In July 1974, a sixteen-year-old Midwestern farm boy who had recently been treated for Hodgkin disease presented to his doctors with marked difficulty breathing. A chest X-ray showed widespread infiltration of the air spaces in his lungs with infected fluids. None of the therapies attempted relieved the boy's progressive respiratory distress, and he died about two months later. On his X-ray both lungs were "whited out"; nearly all of the alveoli were filled with inflammatory fluids and blood. At autopsy, the pathologists also found bloody collections in his pleural cavities, the spaces bordered by the diaphanous membranes surrounding the lungs, which are normally filled with only small amounts of clear cushioning fluid.

Virus was isolated from the boy's lung tissue. It grew readily in embryonated hens' eggs and in cell cultures consisting of rhesus monkey kidney cells. The virus isolate was also found to bind to guinea pig red blood cells (more about all these seemingly odd behaviors later). To further elucidate the morphology and origins of this respiratory virus, a sample was submitted to the Centers for Disease Control and Prevention in Atlanta. Under the lens of an electron microscope, the culprit looked like influenza virus, and its activity could be neutralized by antibodies that had been raised to an earlier flu strain, A/Swine/Wisconsin/67 (aka Hsw1 N1). Oddly, the farm boy's virus was the only influenza isolate among scores of viral species that had been isolated in the Midwest during the summer and mid fall in the previous eight years. (In the United States and other temperate climates, influenza viruses normally reach their peak prevalence in the coldest months of the year.)

Information obtained from the young patient's family revealed that the diligent boy performed chores involved in raising swine on the family farm. He had had daily contact with the animals up until five days before he died. It was hypothesized that the boy's recent chemotherapy treatments for Hodgkin lymphoma dampened his immune system's capacity to respond satisfactorily to the flu virus's fury.

Two of the breeder sows on the farm were found to have antibodies in their blood to the Hsw1 N1 virus strain, indicating that they had been infected at some previous time, possibly very recently. Even more interesting, both of the sows, the boy, and the boy's parents all had antibodies to an even earlier, unrelated epidemic strain of flu, A/Hong Kong/68. It's clear that influenza infections in this tight-knit farm cluster, comprising humans and animals alike, were not unaccustomed to spreading from host to host, whenever close contact took place.

Influenza has often been called the last great uncontrolled plague of man. In the United States, during epidemic periods, it is the only infectious agent that predictably causes a marked spike in the overall national death rate. During some years it smolders as a nuisance, but every few years it arouses worldwide terror and generally defies control by immunization, which is always less than perfect. Surprisingly, unlike the virus itself, which constantly changes, a typical bout of influenza in the typical patient has not varied in its expression for at least the last several centuries, probably more.

The influenza virus has three unique properties not shared with any other respiratory pathogen: It has an unusual capacity for antigenic (protein structure) change; it has the capacity to produce pandemics; and it has an unusual proclivity for causing pneumonia and death. Basically, then, we have an unvarying disease, caused by a virus that varies unpredictably and is unique among infectious agents in its capacity to change its outer protein structure from year to year. These structural alterations allow the virus to evade the full immunologic arsenal that we carry inside us from year to year or from one decade to the next, an arsenal boosted either by natural infection or vaccination.

The influenza virus replicates in epithelial cells quite efficiently, it does not hang around for very long, and it does not remain latent for years, unlike herpes viruses (for example, Herpes simplex and Epstein-Barr virus). Usually, it causes acute inflammation of the inner lining of the trachea, together with fever, prostration, cough, headache, and backache. Bona fide influenza is not merely a "cold" or stomach bug. Influenza generally comes on fast, knocks us out, and sends us to our beds in miserable defeat.

Because of its alarming ability to spread from person to person by the respiratory route, it is not at all surprising that influenza finds easy prey among military recruits crammed together in crowded barracks. And so it happened: In January 1976, a nineteen-year-old Army private stationed at Fort Dix, New Jersey, joined his buddies on a fifty-mile training hike. When he started, he was already ill and defied his commander's orders to stay behind. Before the hike ended, Private Lewis collapsed and died. He was diagnosed as having influenza, and the results of diagnostic testing showed that the responsible virus belonged to the swine flu family of viruses. This novel virus was named A/New Jersey/76 (Hsw1 N1). (Note the striking resemblance to the name assigned to the virus that was isolated from the Midwestern farm boy eighteen months earlier.) The similarity proved that the same virus strain was sustained in the pig and human population, albeit without major clinical consequences, for at least eighteen months.

In January 2006, thirty years after Lewis' death, the CDC published a detailed retrospective account of all ensuing events at Fort Dix. At the time of the recruit's death from the flu, there were 19,000 people living at Fort Dix. A third of them were basic trainees who had arrived at the base about eight weeks earlier. After thorough examinations and indoctrinations the recruits were divided into fifty-member platoons. In order to prevent outbreaks of respiratory illness, recruits were isolated from one another in companies of four platoons each. They had close contact with comrades in their own platoons, less contact with other trainees in their company, and minimal contact with recruits in other companies.

Upon arriving in the autumn for training, all recruits were immunized with a vaccine that contained a 1974 vintage A/H3N2 strain, a different 1972 vintage A/H3N2 strain, and a type B strain that originated in Hong Kong in 1972. Of civilians living on the base, only forty percent agreed to be vaccinated.

All training on the base stopped during the Christmas holidays, and vacationing recruits returned to the base from leave in their various hometowns on January 5, 1976. It was very cold, with wind-chills as low as −43°F. Reception areas were very crowded, and an explosive outbreak of respiratory disease was documented and was found to be due to adenovirus type 21, a serious pathogen well known for its wanton spread among military recruits and for which there is no vaccine. But, so as not to leave any stone unturned, epidemiologic surveillance and testing for other viruses continued. Among the viral cultures obtained and examined at the CDC, four different influenza viruses were identified soon thereafter: One was similar to a strain contained in the vaccine the recruits had received before their winter holidays; the three other strains were previously unknown. In addition to the recruit who died of influenza in January, a few other recruits got very sick during the early part of February. After extensive testing, the soldiers were all found to be infected with a novel strain of influenza, Hsw1 N1, one that had never previously been isolated at the New Jersey Department of Health Virus Laboratory. This alarming discovery was confirmed at the Walter Reed Army Institute of Research. Blood from the diseased recruits was also found to contain antibodies to Hsw1 N1, indicating that, despite their capacity to mount an immunologic response to the virus, the virus won out in the end. These same antibodies also neutralized the flu virus isolated from the lung of the Midwestern pig farmer whom we met earlier, demonstrating that the Wisconsin and

New Jersey influenza strains were identical. Even more terrifying was the recognition that both strains also resembled the H1N1 strain that killed over 40 million people worldwide in 1917–18. When ninety-five additional basic trainees showed up in the hospital or emergency room at Fort Dix with acute febrile, respiratory disease, all hell broke loose; the United States was soon hurled into panic mode, and the US Treasury was about to be depleted of billions of dollars set aside for research, public health interventions, and legal fees.

The time has come for a segue into a brief virology lesson, without which you're not likely to fully appreciate the splendid intricacies of influenza virus structure and function. For the purposes of this lesson I will take the liberty of referring to viruses as microorganisms, that is, living creatures. It's a matter of eternal debate whether viruses should be called living things. They're not capable of locomotion, and they cannot exist independently, that is, outside the confines of living plant or animal cells, except for brief periods of time (sometimes for as long as a few days or weeks, but only under ideal conditions). They are infinitesimally tiny; influenza viruses range in size from fifty to one hundred nanometers (nm) in diameter, or approximately one one-thousandth the diameter of the average human hair (or fifty one-thousandths the size of a one-micron-wide computer circuit). Their magnificent symmetries can be visualized only under the lens of an electron microscope.

While the genetic machinery of animals, including humans, consists entirely of double-stranded DNA molecules collected into units called genes, the genes of viruses may consist of either DNA or RNA; both are nucleic acids. Sometimes the RNA or DNA is a single-stranded molecule and sometimes it's double-stranded. Double-stranded RNA viruses are rare. The primary classification system for viruses is based on the nucleic acid composition of each virus and the number of nucleotides, or genetic subunits, in each virion. A second taxonomic tier is based on the virus's physical structure. Some viruses are surrounded by an envelope and some are not. Influenza viruses are enveloped by a filmy cloak, a combination of proteins and lipids. Such is the face, the contour, that the virus first reveals to its environment, and its structural components determine how the host responds defensively to this potentially treacherous invader; antibodies are most often this first line of defense. The envelope is also the structure that makes contact with and later clutches the cell membrane of its prey.

At the core of most viruses lie single RNA or DNA molecules, that is, the viral genomes which exist as threads of varying length. In others, the viral genome is broken into pieces; such is the case with influenza. The influenza virus genome is divided into eight segments. Each one essentially acts as a single gene and therefore encodes a single protein. Because the eight segments are able to move about independently of one another, should two different influenza viruses happen to invade the same target cell, the sixteen pieces can reassort during the replicative cycle, producing progeny viruses carrying a mix of eight pieces of each parental strain. The caveat is that each infant virion may have *only eight pieces* of RNA, each one responsible for encoding an essential protein (some structural, some functional) that a virus needs to fulfill its destiny—that is, to invade another cell and make more copies of itself, and so on.

Two of the eight proteins give each flu strain a major portion of the identity by which it is universally known; these proteins are the hemagglutinin (H) and the neuraminidase (N). The hemagglutinin renders the virus capable of attaching to a target protein on the mammalian or avian host cell or to proteins on the surface of red blood cells from a handful of species, such as guinea pigs, or on the kidney cells of rhesus monkeys. (These attachment characteristics serve as the basis for the classic diagnostic tests used to identify flu viruses.) The neuraminidase allows the intracellular progeny virions to create an opening in the host cell's membrane so they can run free in fully mature form, while simultaneously taking a small piece of the host cell's membrane to help construct its own envelope. Thus, each and every fully mature influenza virion bears a signature, a pirated piece of the parasitized cell that gave it birth. It also takes with it some hemagglutinin and neuraminidase molecules that it just recently created. What a clever way to give oneself new life, even if you're not really living, at least by the strict criteria assigned to living things by traditional biologists.

The latest swine flu strain to cause a huge fuss in the world, A/California/04/2009, was a veritable Frankenstein's monster. The hemagglutinin came from an already existing H1N2 swine flu variant, the neuraminidase from a Eurasian H1N1/H3N2 double reassortant swine flu strain, and the remaining six RNA segments from three sources: a swine flu triple reassortant that bore genes originating in a North American avian population, a classic swine virus, and a human strain designated H3N2. The human H3N2 prototype originated in Hong Kong and caused a horrific worldwide pandemic in 1968. Subsequent variants of the Hong Kong strain were identified in England in 1972, in Port Chalmers, New Zealand, in 1973, in Scotland in 1974, and in Texas in 1977. Since various H3N2 descendants continue to circulate to this day, an H3N2 variant has been a frequent component of every influenza virus vaccine licensed in the United States for the past several decades. As a consequence, those of us who have been infected with wild influenza viruses or been inoculated with various iterations of the annual flu vaccine have received into our bodies pieces of live attenuated or killed viruses that can boast animal and human forebears from all over the world.

Influenza virus RNA mutates at a rapid rate, thereby changing the biologic characteristics of the virus from one year to the next. Major alterations in either the hemagglutinin or neuraminidase genes result in significant switches in the proteins, or antigens, they encode; this is known as "antigenic shift." Each such shift is almost always accompanied by a major epidemic and sometimes a pandemic, such as those that occurred in 1918, 1929, 1946, 1957, 1968, and 2009. Only one major variant circulates in the population at a time and is ultimately superseded by another. Each variant is assigned one or two new numerals that follow the H and the N in the assigned designation. These numerals mark the place in the sequence of changes that have already occurred in the H and N proteins (for example, H1N1 or H3N2). In interpandemic periods, additional recognizable variations, albeit relatively minor ones, occur at about two- to three-year intervals. These smaller changes in the hemagglutinin or neuraminidase are referred to as "antigenic drift," and account for the geographic moniker that we assign to each "drifted" strain (for example, Hong Kong; Port Chalmers). These little drifts almost certainly result from minor mutations in the original prototype strain, that is, the one that caused the last pandemic.

Back at Fort Dix, an extensive investigation was begun, commandeering the talents of epidemiologists, virologists, and laboratorians. Widespread serologic testing began—it was critical to find out who on the base had antibodies to the swine virus hemagglutinin, a certain marker of prior or current infection. This tactic was very useful in identifying newly infected trainees, mostly older teens or young adults, that is, persons who were unlikely to have had contact with any prior wild flu virus or an older vaccine carrying the swine hemagglutinin. Serologic testing would also identify older soldiers and residents on the base who were not necessarily infected with the new swine flu but who had been infected or vaccinated in years past with flu viruses bearing some older swine hemagglutinin spike. In addition, laboratory facilities were put in place manned by technicians who were expert in isolating and culturing the virus in cell culture from those soldiers who appeared at the infirmary with acute febrile respiratory illnesses. Of 95 throats cultured, A/Victoria/75, an older H3N2 virus, was isolated from half, but A/New Jersey/76, the new Hsw1 N1 strain, was *not* isolated from even one soldier's throat! One can only imagine the number of question marks that rose unexpectedly over the weary heads of all involved.

Despite these surprises, it was critical that the outbreak be fully investigated, so the studies continued. The techs were now engaged in even more extensive analyses of both stored frozen and fresh serum specimens from labs at Fort Meade, Maryland, and Walter Reed in Washington, D.C. Luckily, one worry was quickly extinguished along with a question mark. It was only rarely, on the order of 0 to 8 percent, that antibodies to A/Victoria/75 cross-reacted with antibodies to Hsw1 N1 or to other closely related H1N1 strains. This meant that there was more than a 90 percent chance that any soldier who had antibodies to the current swine flu strain was truly infected with swine flu and did not carry those antibodies as a result of prior vaccination or of a previous infection with wild A/Victoria. After all was said and done, only thirteen enlisted soldiers, ages 17 to 21 years, were discovered to have sustained infection with Hsw1 N1 influenza that was severe enough to require hospitalization. All of those got sick between January 12 and February 8. As we know, one of the thirteen recruits died during a routine march. At autopsy, his lungs were filled with inflammatory fluids replete with white cells and blood. There was no evidence of any concurrent bacterial pneumonia, as so often occurs in fatal cases of flu.

Of the two companies with infected patients, all came from a mere seven platoons. Except for those housed in the same unit, none of the other hospitalized patients knew one another and none had had contact with swine in the six months before they arrived at Fort Dix. There was no evidence of rapid spread of infection and no evidence that the majority of infections resulted in manifest disease or hospitalization. Some members of the seven platoons had antibodies to the swine flu, but the average number of *symptomatic* soldiers was only twenty-five percent in each affected platoon. Even during the week in which the greatest number of flu admissions to the hospital occurred, only about three percent, or twelve, of the trainees in the affected companies needed to be hospitalized.

In outbreaks like this one, even though it was limited and never reached epidemic dimensions, it's critical to know when it's over. One wants to know when active surveillance and intensive testing, both of which are very costly, can be scaled down. It's also good to know when the disease is less likely to spread beyond the confines of "ground zero," so the public can relax. Those basic trainees who were definitively proved to have swine flu had arrived at Fort Dix between January 12 and 26. None had arrived later, so there seemed little chance that any more recent arrivals were carrying swine flu from a source beyond the base. Just to be sure, on February 21 and again on February 27, soldiers from affected companies and thirty-nine soldiers at the reception center were tested for antibodies to Hsw1 N1. None carried antibodies on either occasion, meaning that they did not come to Fort Dix already infected and did not acquire swine flu in the intervening six days. (The incubation period for influenza is just two to three days.) The news was getting better and better.

Unfortunately, there was just one additional cause for alarm. Epidemiologists were temporarily unnerved when they learned that seventeen percent to forty-four percent of civilians on the base had antibodies to swine flu, but their initial concerns were allayed when it was discovered that all the seropositive individuals were more than fifty-one years of age. It seemed very likely that these old-timers had acquired their antibodies years earlier, in response to some long-past outbreak or to some previous exposure to a vaccine containing swine flu antigen.

Finally, by the end of February 1976, influenza A/New Jersey exited Fort Dix for good, but only after 142 recruits were proved definitively to have been infected. These recruits not only had antibodies to swine flu, but influenza virus was actually isolated from the respiratory secretions of each and every one of them. Happily, many of them were not even sick enough to seek medical attention. As we've learned, influenza, like most viruses, does not cause mayhem in every person infected.

Through the middle of March, extensive surveillance for influenza A/New Jersey was carried out in hospitals, clinics, and doctors' offices in a wide geographic swath surrounding Fort Dix. Influenza virus was isolated from 301 people, including soldiers at McGuire Air Force Base (close to Fort Dix), Lakehurst Naval Training Center, and as far away as Dover Air Force Base in Delaware. All the isolates were influenza A/Victoria. No influenza viruses of any lineage were isolated in New Jersey later than March 19. Influenza A/New Jersey/76 was never isolated outside Fort Dix. Hsw1 was not a virulent strain. Its current pattern of poor transmissibility indicated that it did not resemble the devastating H1N1 strain that decimated the world in 1918.

The flu outbreak at Fort Dix was anomalous because of its short duration, its limited scope, and especially the mystery of its origin. How did a farm animal pathogen, well known in many of its prior incarnations, make its way into a hardy group of young trainees, none of whom reported any recent contact with pigs? It seemed even odder, at first glance, that this infection, prone to epidemic potential, did not spread beyond the basic trainees on the base. It was later reasoned that contact between trainees and civilians was limited and that many of the civilians had likely been vaccinated with multiple previous polyvalent, that is, combination flu vaccines, some of which were known to contain swine flu antigen. Many persons on the base were shown to be infected with influenza A/Victoria, and it was hypothesized that perhaps immunity to A/Victoria limited the impact of A/New Jersey or even interfered with the capacity of A/New Jersey to invade its favored target cells in the trachea. This interference phenomenon has not been proven among members of the respiratory virus family group, as it has for some gastrointestinal viruses, but as with multitudes of biologic phenomena, the great majority remain unexplained. There are many intriguing hypotheses that cannot easily be explored.

Inquisitive external observers and rationalists would probably have advised public health sanitarians and government officials to take a deep breath when news broke that the new swine flu was not rampaging outside Fort Dix. So it was surprising that on March 13, 1976, the director of the CDC requested funds to develop and test a new vaccine that could effectively combat A/New Jersey/76. The CDC was asked to plan to produce 80 million doses, enough to stop an expected epidemic. The gears in Washington began to spin and the Secretary of Health, Education, and Welfare (now Health and Human Services, HHS) forwarded the request to President Gerald Ford. The president convened a small meeting of the country's most eminent influenza researchers, who all agreed that A/New Jersey had the potential to cause a pandemic not unlike the 1918 world-wide plague. These experts estimated that the likelihood of a worldwide outbreak was one in ten.

This could potentially lead to 56 million cases of swine flu, with an associated death rate of one per thousand. President Ford called a press conference, and with the polio vaccine icons Jonas Salk and Albert Sabin at his side, he authorized $135 million to be used to develop a swine flu vaccine and to immunize the country's citizens. Other than persons in the Midwest and in areas surrounding New Jersey, this emergency announcement came as a frightening surprise; many of the nation's citizens had never heard of swine flu. The president signed an authorization bill on April 15 (but not before he made a terrible gaffe by announcing that A/New Jersey/76 was identical to the influenza virus strain that killed tens of millions of people worldwide in 1918, including 500,000 in the United States). Immunizations were slated to begin in October.

Although the World Health Organization did not follow suit, American epidemiologists and virologists were vigilant in following up on their investigative work. They determined that the Fort Dix virus was much less virulent than the 1918 flu, and that it had not and almost certainly would not continue to spread, vaccine or no vaccine. The conditions that existed in 1917–18, when the world was engaged in a terrible war with millions forced into international deployments and thousands crowded together in muddy trenches, did not exist in 1976.

Hard to believe, but things really began to go downhill in the late summer and fall of 1976. Two members of the American Legion, attending a conference, died of a fulminant respiratory infection in Philadelphia. Were they the sentinel victims of the re-emergence of the dreaded swine flu? It took the CDC months to determine that the two men had died of atypical pneumonia, now called Legionnaire's disease, which was caused by a previously unrecognized bacterium, not a virus. It was not swine flu. Nevertheless, based on the fear that it might have been, panic ensued, and for at least a few weeks, it set back the swine flu vaccination program. The program was further set back, quite predictably in our ever-litigious society, when the drug companies charged with producing the vaccine in a ridiculously short time refused to proceed until they were thoroughly indemnified by Congress against all lawsuits emanating from the pending vaccine roll-out. Litigation would have cost millions of dollars, even in 1976.

The nationwide swine flu immunization program began on October 1, 1976. Over the next ten days, 40 million Americans received the vaccine. I was one among this patriotic crew of vaccine recipients. I remember standing in line at Yale-New Haven Hospital for about an hour. As soon as I smelled the alcohol used to cleanse my deltoid, it revived a Proustian memory of having previously stood in line in the auditorium of P.S. 182 in East New York, Brooklyn in 1955, to be one of the luckiest first batch of kids to receive my first dose of the Salk polio vaccine. I've always been good about getting all my shots. Somehow, I had learned that vaccines, along with antibiotics, were among the most miraculous inventions of all time and that they have saved many millions of lives. I learned later that the swine flu vaccine was not one of these medical miracles.

The Legionnaire's outbreak was solved and the swine flu immunization program was in full but disappointingly sluggish progress, when the death knell sounded the demise of the infant swine flu vaccine. Suddenly, reports began to appear of the occurrence of Guillain-Barré syndrome in recent recipients of the vaccine. GBS is a very serious and sometimes fatal disease, characterized by an ominous ascending paralysis that results in respiratory arrest when the disease reaches the level of the diaphragm. Some patients with GBS are ultimately connected to ventilators. With so little warning, there was no possible way to definitively connect cases of GBS to prior receipt of flu vaccine, but it was a good enough suspect, particularly because GBS has often been associated, as least temporally, with prior viral infection. The American public could not now be cajoled, let alone convinced, to take the vaccine. First, there was no evidence that swine flu was ravaging the country. Second, who would choose to receive a vaccine that presented a risk, even a small risk, of paralysis or death? After 45 million doses of the swine flu vaccine were distributed, only half the hoped-for number, the entire program was ditched in December. The multiple etiologies and pathogenesis of GBS are still incompletely worked out.

Years passed, and, not surprisingly, swine flu reared its ugly porcine head again in 1988. In early September of that year a previously healthy thirty-two-year-old woman in her thirty-sixth week of pregnancy attended a Wisconsin county fair with her husband. They were particularly interested in spending time in the pig barns. On September 6, the man developed classic signs of influenza—cough, headache, generalized aches, and fever. The following day his wife became ill with the same cluster of symptoms. Unlike her husband, her cough worsened; she experienced shortness of breath and stabbing chest pain, especially when she took a deep breath. This latter symptom is a classic sign of pleural inflammation, sometimes known as pleurisy. The pleura adhere medially to the outermost borders of the lung and laterally to the innermost surfaces of the chest wall, immediately proximate to the ribs and chest muscles. When the naturally clear, cushioning layer of fluid between the pleural membranes gets inflamed, it gets thick and sticky as it fills with endless numbers of the white blood cells that quickly invade to do battle with the marauding viruses or bacteria. Fibrinous strands of newly created scar tissue also grow like ugly webs. For the hapless victim of flu complicated by pleurisy, each deep breath is accompanied by the uneasy scratching and yanking of the pleural membranes against one another and against the otherwise free and easily mobile chest wall. The associated pain can be unbearable.

Our pregnant patient's unremitting cough, chest pain, and dyspnea persisted for the next six days. On the sixth day she visited her obstetrician, who was alarmed to find that her lips and fingertips were bluish and that her lungs were filled with the fine crackling sounds of rales, not unlike the sound produced by a crumpled sheet of cellophane as it scrunches through your fingers. The next day she was admitted to the hospital, but her deteriorating condition demanded transfer to a better-equipped regional referral center. Soon thereafter she was placed on a ventilator and the next day she gave birth to a healthy baby. Four days later, however, despite heroic efforts of the highest order, she died of respiratory failure; her lungs simply could not function any longer. At autopsy they were abnormally heavy; the air spaces were found to be almost entirely filled with fluid, as were both pleural cavities. An aspirate of tracheal secretions collected on the day of admission to hospital grew influenza A, and her antibody titers rose sixteen-fold in the space of seven days. However, this robust antibody response was incapable of halting the relentless arc of influenza's damage. The patient's lung and serum samples were sent to the CDC, and it was discovered that the patient's viral isolate was closely related to an isolate recovered from a sick Wisconsin pig one month earlier.

It did not take a great leap of sleuthing to arrive at the conclusion that this unfortunate woman and her husband were infected during their visit to the pig barns at the county fair. Although half the swine farmers and veterinarians at the fair had had influenza-like illnesses in the first week of September, investigators could not firmly exclude the possibility that the turkeys at the fair were responsible for inciting the outbreak. Sometimes birds and swine share or exchange flu virus genes, particularly when they all attend the same county fair. But because swine seemed to be the most likely vectors of this newest outbreak of the flu and because memories of the Fort Dix outbreak and the 1918 pandemic lingered in the minds of epidemiologists and the public, it was deemed imperative to investigate this new outbreak to the point of exhaustion. Were we in for another epidemic or true pandemic? Was it mandatory to go full steam ahead at preparing a vaccine that contained this latest Wisconsin swine flu isolate? Were isolation or quarantine measures necessary? Were healthcare personnel or young children or pregnant women at particularly high risk? Investigators from the CDC, including officers from the Epidemic Intelligence Service (EIS), were sent into the field. They helped to set up local community laboratories, as well as more centralized diagnostic ones.

The initial focus of attention was on the patient's family and close contacts, healthcare personnel, county fair swine exhibitors, and representative residents of surrounding communities. Standardized questionnaires were distributed to ascertain how many persons from these various groups did or did not have "influenza-like illnesses" (fever, cough, myalgia, headache). The self-administered questionnaires were also designed to discover the degree to which the respondents had contact with the deceased patient, members of her family, or pigs at the fair. Due diligence also demanded that information be gathered concerning illness in junior swine exhibitors who had had the most prolonged contact with their animals during the summer months immediately preceding the county fair. Also, what was the intensity and duration of that contact? Those healthcare personnel who had had contact with the index patient or her husband at any time during her illness and hospitalizations were also questioned about the extent and the precise form of contact they had had. In some of these various groups, blood samples were collected to measure antibodies to the Wisconsin flu virus or to any other influenza viruses that had been represented in vaccines that had been widely used in the previous few years. Reports from the initial investigations had a calming effect: there was not an unusual number of influenza-like illnesses in the community during the months of September or October, and influenza virus had not been isolated from any samples in the Wisconsin public health laboratories. Among the dead woman's contacts at home and at work, only her husband had sustained a respiratory illness. He also had high antibody titers to the swine flu virus that could be explained by his contact with the pigs at the county fair and at his job as a pig farm worker in 1984 and, of course, intimate contact with his now deceased wife. The patient's older, surviving daughter had antibodies to swine flu virus, but at low levels.

It was prescient for those health personnel who cared for the farmer's pregnant wife to worry about her and to provide such excellent care. During the same year in which she fell ill and died, the CDC and health departments from multiple states combined efforts to study extensively the greater risk posed by influenza infection to pregnant women. Data were already available to indicate that certain persons were at higher than normal risk for severe influenza and death. Those at high risk included patients with chronic lung disease, such as asthma and COPD; those with neurological disorders or diabetes; and, sometimes, pregnant women.

Of three infected pregnant women first reported by the CDC in the spring of 2009, one died of acute respiratory distress syndrome soon after she was delivered emergently by Caesarean section of an apparently healthy baby while being ventilated mechanically. An influenza virus was retrieved from her nasopharyngeal secretions; it was identified by CDC as a novel H1N1 virus. Of the remaining two women, one became ill a week after her sister was hospitalized with pneumonia; the other, an asthmatic, became ill during the same week as her two sons, ages seven and eleven, presented to a local clinic with severe cough, sore throat, chills, and weakness. The two latter women were prescribed oseltamivir (Tamiflu), a drug used to treat influenza; neither required hospitalization and both recovered uneventfully.

A much more detailed analysis of records from this high risk group was published in April 2010. Of nearly eight hundred pregnant women with laboratory-documented H1N1 infection, sixty-six percent required hospital admission; thirty (or four percent) died. About one-third delivered prematurely, one-quarter needed treatment in an intensive care unit, and twenty percent were ventilated mechanically. It was hypothesized that the common cardiac, respiratory, and immune system changes that accompany pregnancy may also put pregnant women at increased risk of acquiring influenza and sustaining its usual complications. The administration of antiviral therapy to these women was shown to reduce the need for admission to the ICU and death.

Attention was now focused, post-partum, on the deceased mothers' healthcare workers; collectively those workers had about a two-fold greater risk of having elevated antibody titers compared with their unexposed co-workers, but none had had clinically apparent influenza before or after contact with the deceased mother. This observation exemplifies another universal truth: infections can induce robust and prolonged immunity even when they fail to produce clinical disease.

It was a whole different story for the swine exhibitors. Sixty percent of these teenagers had exhibited pigs with a flu-like illness. Three-quarters of exhibitors who were exposed to symptomatic pigs developed elevated antibody titers, but less than twenty percent developed flu-like illnesses. Among all the sick exhibitors and their sick siblings, all developed their initial symptoms within five days of their first exposure to the pigs at the fair. These observations pointed to a serious risk of transmission of infection from sick pigs to healthy youth. It was the observations made during this outbreak that changed our understanding of the high degree of contagiousness of swine influenza when humans and pigs come into close contact.

A study designed by researchers at the Veterinary Medical Research Institute in Ames, Iowa, and published in the *Clinical Infectious Diseases Journal* in 2006 was motivated by a desire to measure the prevalence of swine flu antibodies among various at-risk populations in Iowa, the nation's number one swine-producing state. Blood samples were collected from hundreds of persons at risk, including farmers, workers at meat processing plants, and veterinarians. The swine viruses to which the antibodies were targeted were isolated from animals on farms in Wisconsin, Minnesota, and far away in New Caledonia, Panama, and Nanchang. All three exposed groups of workers carried markedly elevated antibody titers to both H1N1 and H1N2 influenza virus isolates, when compared to unexposed controls. Farmers were by far the most likely to show past exposure to these strains. It was recommended by the study's authors that swine workers be included routinely in future pandemic surveillance operations to identify newly emerging strains, and that designers of future vaccination trials and tests of new antivirals include these high-risk individuals as among the best subjects for evaluation.

It would not be difficult to find a sufficient number of human subjects to include in such trials. The US pork industry employs over half a million persons, and it is estimated that about 100,000 of them work in pig barns that are dangerously overcrowded. Iowans raise about 25 million hogs each year on more than 9,000 farms. While swine flu was once seasonal, as in humans, the infection is now deeply entrenched in animal species and produces disease throughout the year, exposing swine workers in the process.

There were lots of questions: Shouldn't we think about banning the display of sneezing, coughing, grumpy pigs at their time of arrival at state fairs? Shouldn't we keep swine under strict daily surveillance at populous county fairs so that we can easily cohort and quarantine individual pig groupings as soon as one shows flu-like symptoms? Could we invent a vaccine to prevent the dissemination of this virus among pigs all over the world? To address this last idea, two different experimental vaccine prototypes have been created. One molecularly engineered vaccine contained an attenuated H3N2 strain of swine influenza in which one vital protein gene was truncated, thereby damping its pathogenicity. The other vaccine was constructed around an H1N2 core. Unexpectedly, piglets inoculated with this vaccine developed antibodies, but the antibodies appeared to facilitate rather than impede infection with flu viruses of the H1N1 family. At the time of this writing, neither vaccine is ready for widespread use and the work continues.

Thus far, our discourse has focused on two mammals, humans and swine, perhaps creating the false notion that these are the only two groups that are prey to the assaults of influenza. That impression would be false. Influenza viruses are known to infect a broad range of animals, which in many instances share their viral burdens with one another and serve as soup pots in which a hodgepodge of genes and the building blocks of genes freely mix and mutate. The resulting viral progeny can then show off their new outer coats or sequestered novel proteins.

As examples, well-catalogued and well-studied sentinel strains of influenza have been isolated from humans in England, Australia, Singapore, Hong Kong, and New Jersey; from swine in Iowa and Taiwan; from horses in Prague and Miami;

from chickens in Scotland and Kamchatka; from ducks in Ukraine, Italy, and Germany; from turkeys in Wisconsin and Massachusetts; and from puffins in Norway. Among this international mélange of viruses, the hemagglutinin has tended to change much more frequently than the neuraminidase. At least nine different hemagglutinin iterations were recorded between 1930 and 1976; some of these variants have disappeared and then reappeared in the subsequent four decades, mostly in birds. The neuraminidase gene seemed less likely to mutate. Only two or three variants appeared in the same 46-year time interval and for decades, the influenza vaccines included neuraminidase markers no higher than N3.

The world was awakened to the clarion call of a new, potentially pandemic influenza virus in March 2013 when China reported the first human case of influenza with an avian strain, H7N9. Quietly, the virus had been undergoing a series of mutations in the last years of the twentieth century into the beginning of the twenty-first. By the end of 2013, one hundred forty-five cases had been identified, resulting in forty-six deaths. Human cases continued to appear through most of 2014, raising the toll to more than four hundred.

Another epidemic began in 2016; it was a bad one, and by the end of the year the cumulative number of cases, including all those going back to the first outbreak three years earlier, was 1,223. The overall mortality rate was thirty percent. Most of the cases were linked to live bird markets. But the H7N9 strain rarely kills the poultry which it so easily infects, and thus far, there has been little evidence of sustained human-to-human transmission. Nevertheless, both domestic and international health watch authorities believe that the H7N9 strain of influenza has the greatest potential for causing a pandemic greater than any other influenza A strains in recent memory. The virus merely needs to gain the capacity to spread readily among humans.

Of all susceptible animals, migrating birds play the most important role in spreading influenza viruses throughout the world, not only among avian species but among mammals, too. For some reason, pigs appear to be the most common mammals to which birds bequeath their viral largesse. Because birds on the go share crowded fresh water or salt water resting spots, and because some duck influenza viruses multiply in the cells of the GI tract and are generously shed into the bathing and drinking water—and also because some flu viruses can subsist in water for days to weeks—migratory birds share their viruses with one another and contaminate large swaths of the Earth, particularly places favored by pigs. Influenza viruses found in shearwaters on the Great Barrier Reef have also been isolated from domestic turkeys in California. A flu virus isolated from turkeys in Massachusetts was found subsequently in mallard ducks in Canada, Wisconsin, and Arkansas. Surface H antigens from some flu viruses of common terns in South Africa are believed to have originated from chickens in Scotland around 1960.

The late Dr. Alfred Evans, eminent professor of public health at Yale and Director of the WHO Serum Reference Bank, composed a potent visual image in 1977:

> My own theory is that the nomadic tribes wandering with their cattle and other animals over the great Mongolian land areas set out by Genghis Khan in the 13th century may be a source of new strains [of influenza]. The close and prolonged intimacy between man, animals, and birds in such a setting would be conducive to the emergence of a hybrid strain. On arrival for provisions in one of the cities like Ulan Bator the new strain might initiate a pandemic that could spread across China to Hong Kong.

For an illustration of just how confounding influenza virus admixture and construction can be, one need only examine the history of one recent influenza virus, A/California 04/2009. First, you must be introduced to the six additional genes, besides the hemagglutinin and neuraminidase, which contribute to the constitution of the complete influenza virus genome. There are PB1 and PB2, PA, NP, the M gene, and the NS gene. These pieces of RNA encode critical proteins that, among other functions, help to build new snips of RNA, ion channels, pieces of the viral coat, and a chaperone that helps guide the RNA to the host cell's nucleus. A/California/04/2009 is a triple reassortant, in that it carries pieces of viral RNA from a trio of parents: The N2 neuraminidase variant and M genes originated in Eurasian bird populations, and found their way into European pigs in the late 1970s and early 1980s and then into Asian pigs in 1993. The hemagglutinin, NP, M, and NS genes were derived from a classic swine flu virus that had circulated in North America for approximately two decades. The remaining two genes, PA and PB2, were traced to an older North American avian influenza virus. Not wishing to cede their claim as obsessive genealogists, the molecular virologists chimed in and sequenced entire viral genomes. They traced the A/California strain to one final grandparent virus, a human H3/N2 strain, which also held the original PB1 RNA segment. Remember, this very H3/N2 progenitor spread throughout the world beginning in 1968 and carries the Hong Kong moniker in deference to its place of original isolation. Some of its related descendants continue to circulate to this day.

The Far East is believed to serve as the grand melting pot where flu-susceptible animals interact most often and where most pandemics arise. Pig farming is a huge industry in southern China, Vietnam, Cambodia, and other countries of Southeast Asia. Migratory birds and domestic farm animals, often swine, share watering holes, muddy shorelines, and the farms themselves. If at times bird excrement does carry freshly-minted flu viruses that replicate in tissues of the GI tract, the contaminated feces can easily foul the water shared with pigs. Since migratory birds fly everywhere and pigs stay put on land while serving as the ultimate vessels in which flu viruses mix and match, it's pretty lucky, indeed, that pigs don't fly.

A week after I sent a draft of the previous pages to the publisher, I found the following story in the Sunday *New York Times*. Entitled "Shelter Cats in the Hot Zone," the article, written by Andy Newman, described an outbreak of influenza caused by the H7N2 strain, a bird flu that allegedly had not been found in any animal in ten years. About five hundred cats housed in New York City's animal shelter system needed to be placed in quarantine. The director of the shelter medicine program at the University of Wisconsin was called in to provide expert guidance. Although this was the biggest flu outbreak ever seen in cats, it appeared to be self-limited and only "slightly transmissible to humans." A gray-striped cat named Alfred was designated "patient zero." No one knew where Alfred picked up N7N2. Robin Brennan, the veterinary director of Animal Care Centers of NYC, told the reporter, "That's to me the scary part. Weird." We may never know where the H7N2 came from, and for me, here is the "scary part"—will it come again in some grander, more ominous guise? Ought we do something to prepare? If so, what?

CHAPTER 3

National Parks Can Be Hazardous to Your Health

HANTAVIRUS

In a paper just published in the peer reviewed journal *Infection, Ecology & Epidemiology*, researchers report discovering the first evidence of Seoul hantavirus (SEOV) in the wild rat population in the Netherlands. The discovery comes on the heels of similar ones in France, Belgium and the United Kingdom in recent years . . .

[According to Ake Lundkvist at the Department of Medical Biochemistry and Microbiology, Zoonosis Science Center at Uppsala University in Sweden] SEOV has been reported in previous research as the only one with potential world-wide distribution; this is due to the ubiquitous occurrence of its hosts, the brown and the black rat, which can be found in urban areas, farmlands as well as nature preserves . . . Lundkvist says that relatively little research has been done on these host rats to date. Reason being: "They are very difficult to trap . . . They are much brighter than other rodents."

Science Daily, February 12, 2015

M*orbidity and Mortality Weekly Reports*, a periodical published by the Centers for Disease Control and Prevention, released a short article in June 1993, describing an outbreak of disease that began a month earlier. Those afflicted reported symptoms including cough, fever, muscle aches, and headaches. It sounded like the flu. Patients lived mostly in Arizona, New Mexico, Utah, and Colorado—the states comprising the Four Corners area of the American Southwest. Many were Native Americans. By the time the report was published, two persons from the same household had already died within days of each other; both succumbed to respiratory failure. Experts initially feared that the disease might be pneumonic plague, a more lethal form of bubonic plague, and an infection well known to cause disease in the American West. By the first week of June 1993, 24 patients fitting the case definition were identified; all had abnormal chest x-rays, including pneumonia and abnormally low levels of oxygen in blood. Half of the patients died within a month's time.

Not surprisingly, the media took hold of the story—not only for its shock value but because this disease affected landscapes that seemed unfamiliar and exotic to most people. Reporters quickly dubbed the outbreak the "Four Corners" epidemic, but the medical community proceeded more cautiously. Blood specimens and tissues sampled at autopsy were quickly analyzed by experts at the CDC. They found elevated antibody levels, indicating recent infection, to the hantavirus group of viruses in the victims' blood. However, the lack of specificity in the test results indicated that this particular hantavirus was previously unknown. What to name it? Four Corners virus seemed like the obvious choice; previous hantavirus isolates had often been named for their place of origin. Objections to the Four Corners moniker were raised by many of the Native Americans who populated the area. They feared that the general public might regard their land and their reservations as being somehow tainted or unapproachable. Interested locals and stakeholders were gathered and tasked with finding a more appropriate and inoffensive name. A number

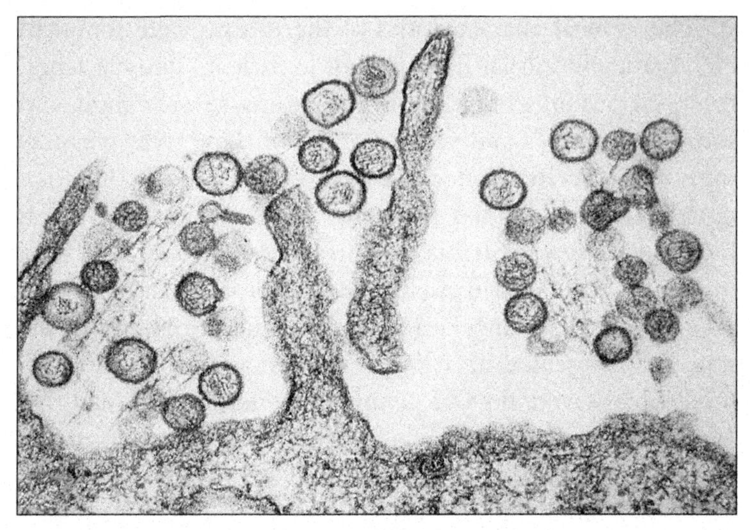

Transmission electron micrograph of the Sin Nombre hantavirus. Hantaviruses cause the hantavirus pulmonary syndrome and are shed primarily in mouse droppings, especially the deer mouse. Hundreds of hantavirus particles have passed through the membrane of an infected cell (at bottom of image) and float free into intercellular spaces or into human secretions or excretions. Image attributed to Cynthia Goldsmith. From the Centers for Disease Control and Prevention's Public Health Image Library (PHIL). Identification No. 1136.

of names were suggested, but none was found to be entirely acceptable. Ultimately, this apparently new hantavirus variant was named Sin nombre, the virus "without a name," or SNV for short.

As news of SNV disease spread, surveillance increased, and within six months of the Four Corners outbreak, scattered cases were confirmed in nine additional states as far east as Texas and Louisiana; no cases were reported from any state farther east. Patients ranged in age from twelve to sixty-nine years, a little more than half were Native Americans, and sixty-two percent of case patients died. The deer mouse was identified as the principal host animal and the presumed disease vector.

The clinical characteristics of the disease were found to be remarkably similar from patient to patient—muscle aches, fever, various upper respiratory symptoms, severe cough, and sometimes nausea and vomiting. All of these were frustratingly non-specific, indeed very much "flu-like." But then quite unexpectedly and suddenly, acute respiratory distress took hold in some patients, an outcome less seldom seen with influenza. One of the patients studied most exhaustively was a 19-year-old New Mexican who lived in a rural part of the state. He presented at a local emergency department with moderately severe flu-like symptoms—nothing unusual. He was a marathon runner who had previously been in excellent health. He lived with his fiancée. She had died two days earlier of a rapidly progressive respiratory disease. The young man had a temperature of 103 degrees and was breathing a little fast. His chest X-ray was normal and his blood tests looked pretty good, so he was sent home with prescriptions for two antibiotics, one for community-acquired pneumonia and one for influenza. He returned two days later with a lower than normal body temperature, low blood pressure, and complaints of vomiting and diarrhea. His kidney function was normal. The remainder of his physical exam was within normal limits, so he was sent home, only to return soon thereafter with a terrible cough producing copious amounts of yellow sputum tinged with blood. His breathing became more and more labored, and he suffered acute respiratory failure and cardiac arrest. He could not be resuscitated. As with this unlucky young man, once hospitalized, most patients in this outbreak developed hematologic disturbances and rapid decline in lung function. Shock was a common accompaniment. Postmortem exams were performed on many of the deceased in an attempt to learn as much as possible about the more sinister aspects of this novel disease. The lungs were heavy with fluid that had displaced much of the air; there were effusions in the pleural cavities. White blood cells, the body's infection warriors, infiltrated the thin septations between the alveoli. Surprisingly, many large mononuclear white cells were found in great abundance in the spleen and liver, indicating that this hantavirus infection was not entirely limited to the lungs; it was a systemic disease, which was later confirmed when viral proteins were found in endothelial cells, those lining the interior of blood vessels, in every organ examined.

The word "hantavirus" remained largely beyond the ken of most Americans for nearly twenty years, but suddenly the name of this renegade was on the lips of many people when, on August 29, 2012, the CDC, by way of the National Park Service (NPS), announced that there were three confirmed cases of hantavirus pulmonary syndrome (HPS) in visitors to Yosemite National Park who had stayed at Curry Village, a collection of canvas tent-cabins on platforms in the Yosemite Valley. At the time of the announcement, two people had already died of the disease. The Park began the arduous task of contacting all persons who had stayed at Curry Village from mid-June through the date of the announcement. The CDC, the California Department of Public Health (CDPH), and the NPS worked together to respond to the situation. Yosemite set up an emergency hot line to respond to questions regarding hantavirus infection. The CDC did the same. Simultaneously the CDPH and the park staff conducted surveys for rodents with the goal of ascertaining the approximate numbers of deer mice in the park and determining the extent of hantavirus infection in the mouse population. A national health advisory was disseminated to all health providers in the country alerting them to the possibility that HPS should be considered in patients with respiratory disease who might have been exposed to rodents recently or who had been to Yosemite during the summer of 2012.

By early September, there were eight confirmed cases of HPS, including three deaths, among visitors to Yosemite. Seven of the eight had stayed at Curry Village, but the eighth victim had hiked in Tuolumne Meadows and the adjoining High Sierra camps, about fifteen miles from Curry Village. Clearly, the infected mice were everywhere! By November 1, ten cases had been confirmed. The CDC continued to support the NPS response by testing patient samples and by providing guidance to healthcare personnel on the clinical manifestations and management of HPS. The CDC also began to search for possible instances of person-to-person transmission, "contact-tracing."

At the time of the outbreak, there were two mouse species, the deer mouse and the white-footed mouse, and two rat species, the cotton rat and the rice rat, that were known to carry hantavirus in the United States. The preferred range of habitation for the cotton rat varies from woodlands and deserts to areas with shrubs and tall grasses. In contrast, the rice rat carries the Bayou hantavirus and prefers marshy and semi-aquatic terrain. Investigators felt obliged to begin sampling blood from all these animals to learn to what extent the hantavirus might have already jumped species and ecological niches.

The hantaviruses are members of what most experts believe is the largest family of related animal viruses, the Bunyaviridae. The more than three hundred members of the family cause a range of human diseases, some quite mild and some more severe, with high mortality rates. The hantavirus name derives from the Hantaan River in Korea, in which vicinity an outbreak of hemorrhagic fever occurred during the Korean War in the early 1950s. Several thousand soldiers developed serious infections, which, in many, included renal failure, hemorrhage, and shock. However, it was not until 1978 that the Hantaan virus was isolated from the lungs of asymptomatically infected striped field mice.

By the mid-1980s it became clear that dozens of different hantavirus species populated the Earth. Isolated cases of ill-

Cotton rat (*Sigmodon hispidus*). A hantavirus carrier that threatens humans when it leaves its droppings in rural and suburban areas, especially in the southeastern United States, Central and South America. Image is the work of the Centers for Disease Control and Prevention, U.S. Department of Health and Human Services.

ness and small disease clusters were reported regularly in the medical literature and were meant to alert healthcare workers and sanitarians to the possibility that human hantavirus victims might turn up almost anywhere in the world. One such case involved a twenty-one-year-old Glaswegian ambulance driver who, in the summer of 1983, developed fever, pharyngitis, and submandibular swelling due to inflamed lymph nodes. When a rash appeared on his hands and feet, he was prescribed an antibiotic. The young man failed to improve and appeared more toxic; he was hospitalized and received routine supportive care. By the twelfth day of illness his fevers ended, but the skin rash persisted and finally resolved as the affected areas of skin peeled away. Ultimately, the patient's kidneys were the organs most seriously affected, but luckily this aberration resolved spontaneously without the need for dialysis. All blood and urine samples sent for routine culture failed to reveal an offending bacterium or known virus. However, three consecutive blood samples sent to the Special Pathogens Reference Laboratory in Salisbury, England, revealed a significant rise in antibody titer to Hantaan virus. This was a surprise for the doctors in Western Europe, so far from Korea. The patient made a complete recovery.

The Hantaan virus most commonly finds its natural reservoir in field mice in Korea and in the bank vole in parts of Scandinavia. The Scottish ambulance driver must have had contact with some such rodents or their excreta. The nature of that contact was never discovered. Members of the Hantaan virus family appear to affect patients differently depending on where they live. The associated illnesses are more dramatic in the Far East, notably Korea, where renal failure is often severe and sometimes superseded by hemorrhagic features. The European disease, nephropathica epidemica, is milder; the kidneys are less affected and there are no problems with excessive bleeding. The Scottish patient's illness exemplified the less aggressive features of the European flavor of hantavirus infection.

In North America, the first encounter with overt hantavirus infection was the SNV outbreak in the Four Corners area. Because most affected patients sought help due to progressive respiratory distress, the American hantavirus disease was granted the name "hantavirus pulmonary syndrome." The American patients did not have skin rashes, swollen red eyes, or swollen glands, unlike their Far East counterparts.

Like all infections, the associated clinical picture represents a perfect synchrony between the pathogen and host immune response. So there are several ways to explain the varying behavior of hantavirus infection. Perhaps the genetic make-up of ethnic populations living in discrete geographic locales has a different effect on the capacity of the immune system to control hantavirus infection. In some instances, the immunologic reaction may be ineffectual, allowing the virus to replicate overabundantly and to cause limitless bodily harm. In other instances, the immune system may overact, resulting in unexpectedly dramatic inflammatory side effects that may be even worse than the virus's own trouble-making behavior. Or the body's defenses may be "just right," perfectly tuned to the real threat the virus presents and modulated just enough to eradicate the threat and then get out of the way.

If we look at the various manifestations of disease caused by this pathogen, we cannot help but believe that each extant viral strain likely survived eons through adaptive evolution orchestrated by multiple series of mutations. In each geographic niche, the local hantavirus represents the final iteration of a series of changes that allow the local virus to find its perfect partner among the indigenous rodent population and perfect "accidental hosts," often humans, who veer too close to the contaminated rodents and their droppings. It's best if the human host, like the rodent host, does not die too soon, if at all. The virus's modus operandi is to reproduce itself endlessly, and it can only do so when its host stays alive. After all, viruses only replicate within living cells.

Odd stories of hantavirus-human interactions continued to be reported during the two decades following its isolation in the laboratory. Since most human cases are solitary (hantavirus infections are not believed to spread person-to-person) and since the diagnosis is not often considered or proved, these individual cases have a certain "freakish" appeal. The proud discoverer of an oddball infection, such as hantavirus, might even gleefully shout out his find to his astonished labmates.

In that vein, doctors in Nottingham, England, published the case history of a ten-year-old boy who was admitted to the hospital in the spring of 1994 with a gastrointestinal illness characterized by nausea, vomiting, diarrhea, abdominal pain, and blood in the urine. His private doctor prescribed a handful of drugs designed to provide symptomatic relief. Then he discovered on physical exam that the patient had tenderness in the right "loin" (i.e., flank). The boy had no fever, but lab results showed evidence of renal dysfunction, which was supported when an abdominal ultrasound revealed large swollen kidneys. Kidney function quickly deteriorated to such a degree that it was imperative to do a renal biopsy so as to make possible a definitive diagnosis. The pathologists saw patchy infiltrates of white blood cells of all types within the renal tubules. The blood vessels looked normal. Despite a short course of steroids the boy's kidneys began to fail, necessitating dialysis treatments. A repeat biopsy showed even greater numbers of inflammatory cell infiltrates and early signs of scarring.

It turns out the boy's family lived in a "caravan" (i.e., trailer) park adjacent to a "scrapyard" (i.e., junkyard) where the boy and his friends often played. The play areas were known to be infested with rats. The boy's blood samples were sent to a reference lab for Applied Microbiology and Research. There were no antibodies to hantavirus in a sample drawn five days after admission to hospital, but a month later the antibody titers had risen greatly and contained an IgM component, a moiety known to be present only transiently during acute infection. Thirty months after the boy recovered from his illness, his renal function remained compromised, but he was clinically well, with normal blood pressure. The authors reported in their May 1997 publication that their patient "is the youngest so far recorded." They also admonished the reader to consider hantavirus infection in "patients who have acute renal failure due to interstitial nephritis, especially if there is a possibility of exposure to rodents, irrespective of a history of travel to areas of known hantavirus endemicity."

What happens when hantaviruses find their way to the kidney by way of the bloodstream? Much was discovered about the virus-kidney relationship during outbreaks that occurred in the first half of the twentieth century in Manchuria, Siberia, and Korea. The associated disease was named epidemic hemorrhagic fever (EHF), but it was the impact of the virus on the kidneys that soon became an intense subject of interest as a symbol of the more widely dispersed manifestations of the disease. Of the many Americans involved in the Korean conflict, approximately 1,700 fell victim to EHF between 1951 and 1952. As with other diseases that erupt at times of war, such as typhus and pandemic influenza during World War I, there was serious concern about the potential impact on the American effort. As always, this worry evolved into a veritable frenzy of epidemiologic and virologic research, a lot of it orchestrated by service officers assigned to the Eighth Army in Korea and the Far East Command or to the Epidemic Hemorrhagic Fever Field Unit of the Armed Forces Epidemiological Board.

Descriptions of epidemic hemorrhagic fever were uniform in content; the illness began with headache, fever, and chills in addition to anorexia and vomiting. The mid-range fevers generally lasted for four to five days and were accompanied by flushed facies, red eyes, and inflamed mucous membranes. The numbers of white blood cells began to fall, and a spotty petechial rash appeared on the palate and the axillae, a consequence of the sudden decline in the number of circulating platelets, the clotting elements in the blood. Soon thereafter, the decline of kidney function was heralded by a spilling of albumin, the most abundant blood protein, and red blood cells into the urine and, finally, oliguria, a decline in the volume of excreted urine. Many patients began to recover at this point, but others progressed to a critical or fatal end as intractable bleeding into the upper and lower GI tracts, urinary system, and skin ultimately contributed to outright shock. In those who later died, the disease left its ominous stamp in the form of purpura, deep purple, irregularly shaped marks caused by hemorrhage into the skin. For those who didn't die as an immediate result of this bleeding diathesis, some later died of intractable shock brought on by further bleeding into vital internal organs like the adrenal gland, the pituitary, and the heart, or by progressive renal failure.

Careful epidemiologic investigations carried out during these early wartime episodes led to the observation that the disease did not spread from person to person, but clearly had a diabolical transmission cycle that involved rodents. That trombiculid mites might also play some accessory role was posited but later left to rest without further proof.

During the subsequent half century and into the twenty-first, greater and greater attention has been paid to the renal aspects of some of these hantavirus infections, to the extent that some of these, including Songo fever, Korean hemorrhagic fever, and nephropathica epidemica, have been decreed by the World Health Organization to be known as "hemorrhagic fever with renal syndrome" (HFRS), a mouthful indeed—hence the more widely used abbreviation.

Increased vascular permeability, that is, leaky blood vessels, plays an essential role in the pathogenesis of the more severe forms of hantavirus infection, including those targeting the kidney and lungs. Blood plasma and serum proteins leak into interstitial tissues throughout the body. Because the entire vascular space contracts, the elements in the kidney that filter the blood, that is, the glomeruli, function less efficiently, and renal failure ensues. There is evidence to suggest that genetic factors may contribute to the susceptibility to renal involvement, a consequence, perhaps, of magnified immune responses to the virus that result in platelet activation, blood coagulation, and lysis of certain blood proteins.

Some scientists believe that the exuberant immune response may elicit greater degrees of pathology than the virus does itself. This is by no means the only example in which the immune-mediated manifestations of disease exceed the disarray caused by the inciting virus. A good example is infectious mononucleosis (or "mono"), in which the Epstein-Barr virus causes less cell damage and dysfunction than that perpetrated by the abundant, and in some cases the overexaggerated, response of the immune system to viral invasion. The painful, inflamed tonsils, the swollen tender lymph nodes, and the splenomegaly result from the dramatic entry of innumerable battle-ready T-lymphocytes and inflammatory proteins into the various tissues that had previously been infected by EBV. The illness proceeds until the immune system exerts self-control, secure in the evidence that the enemy has been localized and vanquished.

In the case of hantaviruses, renal cells and capillary beds in the lung are infected, but are not necessarily destroyed. However, the infected cells set off a cascade of inflammatory reactions characterized by infiltration of the lung and kidney tissues by lymphocytes, edema fluid, and dilatation of blood vessels. Inflammatory proteins, including antibodies and their antigens, form deposits in the glomeruli.

By the end of the twentieth century, hantavirus infections had come to be recognized in most parts of the world. Physicians needed to understand what specific demographic risk factors might be used to distinguish hantavirus infection from a number of other viral and bacterial illnesses with similar clinical presentations, especially the common ones that occur in the earliest phases of disease. In developing a differential diagnosis, the best-trained caregivers and those with the greatest sleuthing skills question patients about much more than their symptoms and the medicines they take. They also learn something about the patient's household constellation and family medical history. They might ask about recent travel, family pets, work habits, hobbies, dietary preferences, and out-of-the ordinary foodie experiences. If the possibility of hantavirus infection happens to find itself on the list of both common and arcane illnesses that the physician has in mind, such as pneumonic plague, inhalation anthrax, leptospirosis, meningococcemia, and even run-of-the mill community-acquired pneumonia, how does she narrow down the list; what questions does she ask about the details of her patient's work, his living arrangements, or the micro-environments in which he performs both his vocational and leisure activities? Epidemiologists and infectious disease specialists are often among the scientists who contribute most to the knowledge base of practicing healthcare workers who hear the answers to those questions that needed to be asked.

A number of outbreaks of nephropathia epidemica (NE), a mild form of hantavirus renal syndrome, occurred in the Ardennes region of Belgium in the early 1990s. At the time, it was already widely known that Puumala virus was the local hantavirus culprit that caused most of the associated renal disease in Scandinavia and the Low Countries of Western Europe. By the time 62 cases of NE has been reported in southern Belgium, the investigators embarked upon a case-control study that focused on identifying human activities that put persons at risk for falling victim to NE. Case-control studies are among the most powerful epidemiologic methodologies available to investigators. Part of their convenience lies in the fact that they can be done retrospectively, bypassing the enormous expenditures of time and money needed to conduct a prospective study, which can take years and many man-hours to complete. The "cases" in a case-control study are patients identified through medical record review who fulfill very specific clinical criteria, in this case all the signs, symptoms, and laboratory results accepted by the medical community as representing bona fide cases of NE. Then one chooses matched controls, that is, persons *without* NE, usually two or three for each case, who are matched for age, ethnicity, socioeconomic status, and, if possible, similar living arrangements. Questionnaires are administered to the case and control patients that focus on factors that are believed to put one at risk for hantavirus infection. At the foundation of most of the questions asked in the Belgian study lay the foreknowledge that hantaviruses cause infection when inhaled from particles of feces or dried remnants of urine, saliva, or blood of infected rodents. In Belgium, the rodent of particular interest was the red bank vole.

The Belgium study identified fourteen "exposure variables." Merely seeing living rodents at home or in the workplace and, especially, any direct contact with a rodent were highly significant risks. So was the mere sighting of a rodent nest or rodent droppings. Any activity which involved work in the forest or firewood-cutting posed risk. Although reopening a cottage or summer house after the winter did pose an increased risk of hantavirus infection, cleaning the basement, attic, garden house, or woodshed did not. However, the length of time in hours or days one spent performing any of these activities increased markedly the likelihood of risk of infection. For those who developed pulmonary hantavirus infection, tobacco use was found to increase susceptibility to illness. Lots of other activities, for example, picnicking, fruit-gathering, jogging, and fishing, posed no risk at all.

What happens in Belgium doesn't stay in Belgium! The hantavirus outbreak in Yosemite National Park occurred in one of the most heavily forested areas in the US, where visitors camped or hiked and slept in tents and where mice almost certainly spent their winters and springs, built their nests, and raised their families. The excreta they left behind dried and became powdery and likely found its way into aerosols when the tents were prepared for throngs of unsuspecting summer visitors. Sweeping the tent platforms before the tents were occupied by successive groups of visitors probably didn't help matters. The critical epidemiologic and behavioral co-factors associated with acquisition of hantavirus infection soon became part of the routine repertoire of questions that healthcare workers posed to their most puzzling renal or pulmonary patients.

The tents in Yosemite are not unlike the summer houses that were swept clean in Belgium during the outbreaks of disease between 1992 and 1994. The lesson here: beware of freshly sanitized country cottages in Scandinavia, Belgium, and the Netherlands, which lie dormant during those cold, dreary, wet winters. Also, remember that high elevation habitations in the American West can pose a risk to one's health when freshly prepared for the invasion of the outdoorsy summer people.

CHAPTER 4

Prairie Dogs Make Lousy Pets

MONKEYPOX

> [The Centers for Disease Control and Prevention] warned pet owners not to release any sick or potentially exposed animals into the wild . . . A spokesman for the agency acknowledged that the authorities did not know the whereabouts of many of the estimated 850 animals in an April 9 shipment from Ghana to Texas, nor do they know if any were released . . . "That's one of the things we're really worried about," said David Daigle, a spokesman for the agency. "Tracking them all down is darn near impossible."
>
> *The New York Times*, July 3, 2003

May 8, 1980, was a momentous date in the history of medical science and global health. On that day the World Health Assembly announced that smallpox had been eradicated from the face of the Earth. The final two indigenous smallpox infections acquired "in the wild" afflicted a Somali hospital cook in October 1977 and a two-year-old Bangladeshi girl in October 1975. It was predicted that smallpox would never return as an endemic global disease. Prodigious verification activities were undertaken by a cadre of eminent scientists in 1978–79; they did, in fact, conclude that no additional cases of smallpox had been detected in the previous two year interval. So the World Health Organization promulgated the following resolution:

> Having considered the development and results of the global program on smallpox eradication initiated by the WHO in 1958 and intensified since 1967 . . . [the WHO] declares solemnly that the world and its peoples have won freedom from smallpox, which was a most devastating disease sweeping in epidemic form through many countries since earliest time, leaving death, blindness and disfigurement in its wake and which only a decade ago was rampant in Africa, Asia and South America. (World Health Organization, Resolution WHA33.3)

At the time of the pronouncement, the WHO failed to demand that all laboratories working with the virus rid themselves of every remaining stock of smallpox (variola); that came later. When the WHO declared the world free of smallpox, they were referring to a disease that was found "naturally," only among people living in communities of other people and animals. Specifically, they did not mention a case of smallpox that had occurred in a laboratory in Birmingham, England, in 1978. A medical photographer contracted the disease at the city's medical school and ultimately died on September 11. Because he bore ultimate responsibility for the smallpox research performed at this institution, Professor Henry Bedson took his own life soon afterward.

This tragic incident brought immediate attention to concerns regarding the status of safety protocols of other labs in the world that also performed experiments utilizing stocks of smallpox virus. Biosafety quickly became a cause célèbre that ultimately rose to the top of the heap of practices integral to the performance standards of every laboratory that worked with viruses and bacteria, particularly those known to have the propensity to spread rapidly in human populations. Eventually, thoughts of biological warfare and bioterrorism began to gnaw at the conscience of scientists, politicians, and the general citizenry who were familiar with the horrors that might ensue should an agent like smallpox be weaponized.

History provided both allegations and persuasive evidence that smallpox had been used as an agent of destruction during the French and Indian Wars and the American Revolution. Some historians suggested that smallpox had been used by British marines during their subjugation of indigenous aboriginal tribes in New South Wales, Australia, in the last quarter of the eighteenth century. Others documented the pernicious contamination of blankets with smallpox scabs that were given to Native Americans during the Indian Wars of the 1800s. More recently, evidence has been uncovered that scientists from the United Kingdom, the United States, and Japan were engaged in smallpox weapons research during World War II. Luckily, large-scale production of the weaponized virus was halted when it became obvious that the ready availability of an effective smallpox vaccine would abort any effort to bring about deaths on a large scale.

The smallpox virus is only one among a large family of related viruses, some of which scamper among various animal species, one of which is Homo sapiens. The collective memory of the ravages caused by smallpox is responsible for the fear that naturally arises when other poxvirus diseases appear in our fellow humans.

There are four genera of poxviruses, comprising twelve distinct species, some of which are known to cause disease in humans. The majority are distributed worldwide. Unlike some of the others, smallpox infection is limited to humans. Although highly contagious by intimate personal contact, it does not spread either to or from other animals. In its day, smallpox rampaged throughout the world. In contrast, monkeypox virus is native to Central and West Africa; buffalopox to Indonesia, India, and Egypt; and cowpox to Europe and Western Asia.

Vaccinia virus (vacca = cow in Latin), which has traveled all over the world and infected millions of its inhabitants, is the virus which served so magnificently as the active element in the smallpox vaccine, even though it is different from smallpox and despite the fact that its much-storied unique origins from the pocks of a milkmaid and her cows are now widely disputed. (It's possible that some horsepox may have found its way into the preparation concocted by Edward Jenner in 1798.)

The poxviruses are the largest and the most complex of all viruses. They are two to four times larger than influenza viruses and about six to eight times larger than poliovirus, a member of the picorna virus group (pico = small; rna = RNA). The poxvirus genome is comprised of double stranded DNA, which codes for 150–300 proteins, depending on the species. Within each poxvirus genus, antibodies raised by the animal host to one virus are able to neutralize other viral species in the same group. This behavior is the basis for the extraordinary efficacy of the smallpox vaccine. Antibodies raised by humans in response to cowpox, horsepox, or vaccinia virus (all orthopoxviruses) are perfectly able to protect the host who later comes into contact with smallpox.

Genetic recombination occurs efficiently between viruses in the same genus. This particular behavior has the potential for causing trouble. What might happen if a horsepox virus and an imported monkeypox virus were to swap some fragments of their DNA, thereby creating a centaurean hybrid, one that might have a craving for human flesh? That nascent, perhaps irrational fear arose in the fertile minds of some when an odd pocky disease appeared in the United States in the spring of 2003.

The Centers for Disease Control and Prevention and six state and local health departments were called into action

Black-tailed prairie dog (*Cynomys ludovicianus*), housed at the Smithsonian National Zoological Park, Washington, D.C. Photograph by Joe Ravi,

when clusters of victims of monkeypox came to public notice in Wisconsin, Indiana, Illinois, and three other states. All patients had fevers and pustular skin lesions and all had had contact with sick prairie dogs (cynomys species) that were kept as pets. A little gumshoe detective work traced all the sick rodents back to a distributor in northern Illinois who regularly received, housed together, and later distributed exotic African rodents and American prairie dogs throughout the American Midwest.

The first patient who came to medical attention was a three-year-old girl in central Wisconsin who was hospitalized with fever and a soft tissue infection of the hand. She had been bitten on the finger by a prairie dog that was purchased at a "swap meet." The prairie dog developed drippy eyes, swollen glands, and a bumpy rash on May 13, the same day it bit the girl. The animal died seven days later. Thankfully the little girl survived, but only after her doctors successfully ruled out two rare bacterial infections, tularemia and plague, that would have required antibiotics to achieve a cure. On

June 2, the Milwaukee health department was notified of an ailing meat inspector who lived in southeastern Wisconsin. He worked part-time as a distributor of exotic animals; he described the bite and scratch he had received from a prairie dog two weeks earlier. The pus-filled lesion started as a small nodule at the site of the scratch but was soon followed by fevers, sweats, chills, and regional lymph node swelling. He ultimately needed to be hospitalized. It was soon discovered that this seriously ill meat inspector cum exotic animal purveyor had sold two prairie dogs to the parents of the little girl described previously. Once this connection was established, the epidemiologists embarked on a traditional "trace-back" and "trace-forward" operation. By this time the investigators knew what they were dealing with: fluid removed from one of the skin pustules revealed typical poxvirus particles under the electron microscope. Viruses with the identical size and structure were found in supernatant fluids from tissue cultures that had been inoculated with minute amounts of the pus. Finally, and incontrovertibly, the gene sequence of DNA purified from viral cultures of both human and animal origin were found to be identical. The little girl, the meat inspector, and two additional prairie dogs in southeastern Wisconsin all carried the same viral DNA. The pet animals and humans were ultimately linked to a single strange vector, a giant Gambian rat.

Extensive epidemiologic investigations uncovered the full extent and sequence of the monkeypox trek. An animal distributor in northeastern Illinois received a shipment of exotic rodents from Ghana, which included at least one Gambian rat. The imported animals were housed with prairie dogs, and part of this animal menagerie was shipped to another distributor in southeastern Wisconsin. He bought thirty-nine prairie dogs and at least one Gambian giant rat, and he shipped at least two prairie dogs to a family in Wisconsin, three of

Gambian pouched rat cub (*Cricetomys gambianus*). Photograph by Rosa Jay/ Shutterstock.com.

whom fell ill over the following five to twenty days. The distributor and a close contact of his also got sick about three weeks later. The little girl described earlier was a member of the Wisconsin family that had purchased his prairie dogs.

The Wisconsin exotic-pet distributor also shipped twelve affected prairie dogs to a veterinary clinic and two pet stores which thereafter sold infected animals to a family in which two members got sick and to another animal clinic in which a second veterinarian became infected. The investigation of this outbreak proved repeatedly that monkeypox is contagious, but much less deadly than smallpox. It also proved that the importation of rodents from Africa is a dangerous practice and that prairie dogs are sometimes, but certainly not routinely, susceptible to infection with serious pathogens, particularly when purchased from uninformed distributors.

Although some people favor giant Gambian rats as pets because of their "playfulness" and "cuddliness," most folks would probably agree that only a mother could love one. The rodent is also known as an African pouched rat because of the enormous cheek pockets which can temporarily store lots of food for future use. (It's been reported that some of these rats have stuffed their pouches so full of food that they cannot pass through the entries to their own burrows after returning from a food-collecting binge.) Gambian rats are native to all of central Africa, and they are among the largest rodents in the world, regularly attaining lengths of three feet, half of which is comprised of a tail that ends in a bright white band. The rat generally weighs between two and a half and five pounds, so it's about the size of a small cat. Although it's omnivorous, it prefers vegetables and, in Africa, palm fruits and kernels.

Exotic animal importers are responsible for having brought Gambian rats to the United States, and harried pet owners and animal distributors who tire of the rat's voracious appetite and ceaseless activity have been responsible in the past for having let loose those rats who've exasperated them to the point of desperation. In fact, the rats have now found status as an invasive species in the Florida Keys. The rats are quite fertile; they reach childbearing age before their first birthday and have three or four litters a year, with four to six pups in each litter. Should an American rat become infected with monkeypox virus introduced into its habitat by a newly emigrated and liberated African invader, we might be in for a nasty epidemic, beginning with a hotspot in the heat of southern Florida. But maybe not. In 2003, after the Midwestern monkeypox outbreak, the importation of African rodents was banned. (That doesn't mean it couldn't happen.)

Monkeypox is a much less serious infection than smallpox, but it's no picnic. The skin lesions are ugly and tender, and the lymphadenopathy is impressive and equally uncomfortable. The pre-eruptive or prodromal stage is characterized by moderately high fever, chills, headache, sweats, and profound malaise. Some patients have sore throat, cough, and encephalopathic symptoms. The rash, or exanthem, begins about a week after the illness starts, most often at the site of the animal bite or scratch. If the infection is acquired by the oropharyngeal route, for instance by ingesting contaminated meat or by bringing unwashed fingers to the mouth, the rash begins on the face and spreads centrifugally toward the trunk and limbs. Although the vesiculopustular skin lesions resemble those of smallpox and chickenpox, the impressive lymph node involvement clearly distinguishes monkeypox from the

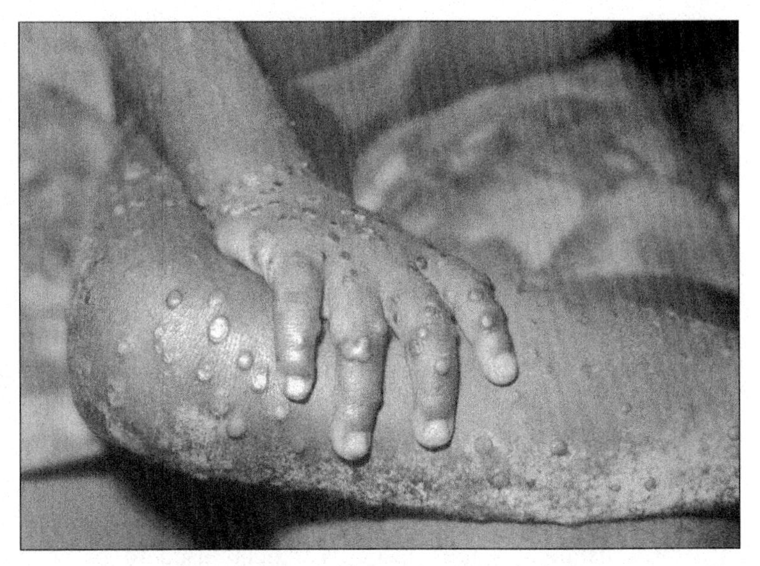

Close-up of monkeypox lesions on the arm and leg of a female child, Grand Gedeh County, Liberia. Humans are infected from contact with the blood or saliva of some rodents and primates. A work of the Centers for Disease Control and Prevention, U.S. Department of Health and Human Services.

other poxvirus diseases. Also, and importantly, when the pocks of monkeypox pop up, they appear and evolve synchronously, unlike chickenpox, in which, at any one time, pocks in the same small area of skin are at different stages of evolution. We would never want an inexperienced clinician to mistake one for the other; we could not risk the panic that would inevitably ensue. Among the more malnourished and immunocompromised peoples of the world's poor countries, the rash of monkeypox may become nearly confluent over some parts of the body. When the pocks finally crust and involute, deep scars remain, which are often unsightly and even permanently deforming.

In the US, monkeypox outbreaks have been linked to only two animals—prairie dogs and Gambian giant rats. In contrast, central Africa is home to many animals that carry the virus; these include the domestic pig, the elephant shrew, and three squirrel species. These animals often end up as food. The virus is easily transmitted from person to person, specifically under conditions of crowding and poor hygiene. There is currently no specific antiviral medication for monkeypox. Fortunately, simple supportive care keeps the death rate low, and the rare death is most often attributed to superimposed secondary bacterial infection.

Whereas, in its heyday, universal smallpox vaccination would have cross-protected against monkeypox, smallpox vaccine is no longer administered routinely anywhere in the world. But a limited number of doses of the vaccine can be found in small reserves, such as those at the CDC. In some rare circumstances a few doses have been used to prevent transmission of disease from monkeypox victims to health-care workers and close household contacts.

Having learned that closely-related orthopoxviruses can swap pieces of their DNA and that the origins of the vaccinia virus smallpox vaccine have been shrouded in mystery for more than two centuries, I will endeavor to clarify the facts surrounding the creation of a smallpox vaccine by Edward Jenner in 1798. I decided to avail myself of a small sample of the treasures displayed and stored within the sacred spaces that house the historical library of the Yale School of Medicine. I wanted to find out whether the smallpox vaccine consisted entirely of infected matter taken from the udders of milk cows and the hands of their milkmaids, as I and generations of medical students and microbiologists had been taught. I was determined not to pass on to my students the same untruths or semi-truths that I was taught quite innocently as a medical student in the years when molecular virology and gene sequencing were sciences that had not yet been invented. The reason for my circumspection lay in the fact that I had only recently learned that vaccinia virus, the actual pathogen found in commercially produced smallpox vaccine, is not identical to either cowpox or smallpox; so what was it? The only thing I did know for sure is that the vaccinia virus, or other viruses so named or otherwise related, which constitute the smallpox vaccine, has saved millions of lives by preventing smallpox over the course of several centuries, at least.

Having been trained in the basic practices of epidemiologic investigation, I set out on a little sleuthing adventure. I strode through three corridors of the medical school, crossed Cedar Street, and passed between the four three-story-high Corinthian columns that form the entrance to the school. I walked through the cool innards of the institution that was founded in 1813 as the Medical Institution of Yale College. I reached the rotunda, an airy two-story oval space with sunlight beaming through open doorways on the second level that bathe the brick-red tile floor and six glass-enclosed wooden display cases in a soft, peaceful glow. I passed through the heavy iron and glass doors on my left and entered the venerable sanctum that is the Medical Historical Library of Yale University. The main reading room is two stories high, and on each level closed bookcases line the walls. The slatted ceiling resembles the bottom of an inverted whaling vessel. Gently lighted reading tables sit side by side along the entire carpeted and wide-planked floor.

The origins of this unique collection of books both ancient and modern are described in various blurbs published online by the University:

> It was the vision of Harvey Cushing, who joined with his two friends and fellow bibliophiles, Arnold C. Klebs and John F. Fulton, in what they called—with many inventive synonyms—

> their "Trinitarian plan," to donate their superb book collections to Yale if Yale would build a place to house them. As the plan matured it became wedded to the idea of creating a new medical library for the Yale University School of Medicine. Cushing was the driving force persuading Yale officials to realize his vision. He wanted the medical library to be the heart of the medical school . . . and that the old and new collections be equally accessible . . . Cushing was informed of the University's approval of the final plans on the day before his death in October 1939. The Yale Medical Library (now the Harvey Cushing/John Hay Whitney Medical Library) was built through 1940 and officially dedicated in 1941.

In an office beside the library reading room, Melissa Grafe, the John R. Bumstead Librarian for Medical History, introduced me to the original publications that described Edward Jenner's discovery and initial experiences with his smallpox vaccines. I was led into a brightly lit room after passing through huge heavy glass doors that could only be opened with an electronic proximity I.D. card reader. The doors closed behind me and Grafe showed me to my assigned place. Jenner's book lay open, its sides gently cradled on two heavy cardboard rests, so that the binding would not be strained. I was told not to move the book to my lap and not to eat or drink. I was permitted to use paper and pencil to take notes. No pens! A laptop was also okay. I was about to embark on a journey that Janice Nimure, in an essay in the *New York Times* in January 2016, called "research rapture"—the moment you "slip the bonds of the present and follow a twinkling detail into the past."

The book I had chosen was *Further Observations on the Variolae Vaccinae or Cowpox* by Edward Jenner, M.D., published in London in 1799.

It was obvious to me in reading the first few pages of the ancient tome that Jenner was determined to prove the ve-

AN

INQUIRY

INTO

THE CAUSES AND EFFECTS

OF

THE VARIOLÆ VACCINÆ,

A DISEASE

DISCOVERED IN SOME OF THE WESTERN COUNTIES OF ENGLAND,

PARTICULARLY

GLOUCESTERSHIRE,

AND KNOWN BY THE NAME OF

THE COW POX.

BY EDWARD JENNER, M. D. F. R. S. &c.

——— QUID NOBIS CERTIUS IPSIS
SENSIBUS ESSE POTEST, QUO VERA AC FALSA NOTEMUS.

LUCRETIUS.

London:

PRINTED, FOR THE AUTHOR,

BY SAMPSON LOW, N°. 7, BERWICK STREET, SOHO:

AND SOLD BY LAW, AVE-MARIA LANE; AND MURRAY AND HIGHLEY, FLEET STREET.

1798.

Title page from Edward Jenner, *An inquiry into the causes and effects of the variolae vaccinae: a disease discovered in some of the western counties of England, particularly Gloucestershire, and known by the name of the cow pox*. London: Printed, for the Author, by Sampson Low, 1798. Courtesy of Yale University, Harvey Cushing/John Hay Whitney Medical Library.

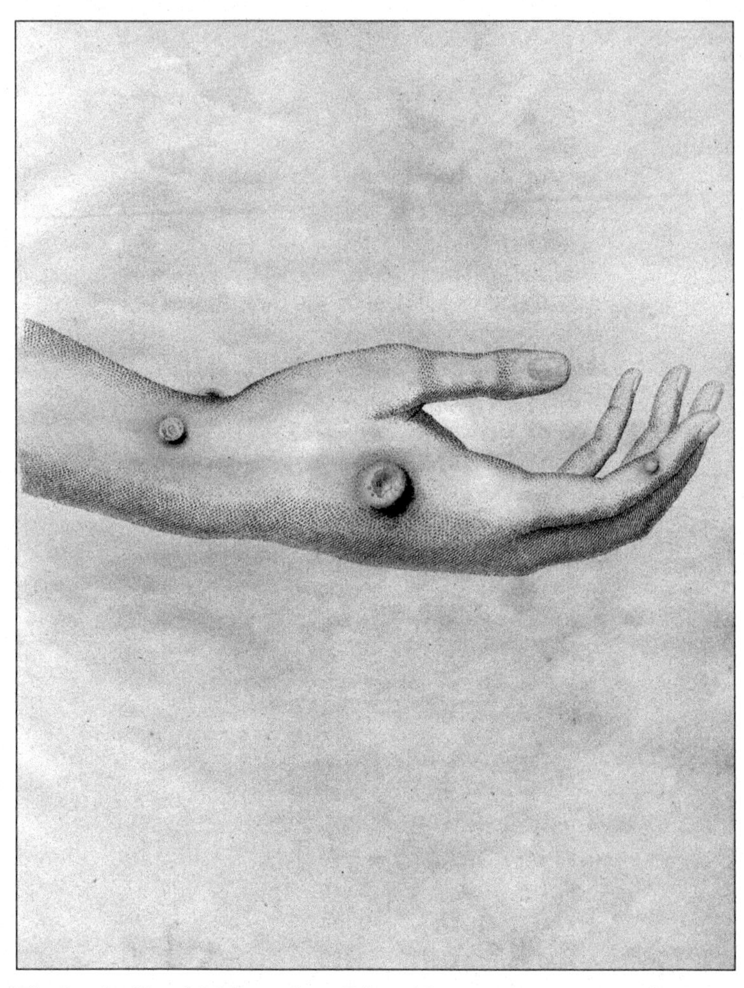

The hand of Sarah Nelmes, from Edward Jenner, *An inquiry into the causes and effects of the variolae vaccinae: a disease discovered in some of the western counties of England, particularly Gloucestershire, and known by the name of the cow pox*. London: Printed, for the Author, by Sampson Low, 1798. Courtesy of Yale University, Harvey Cushing/John Hay Whitney Medical Library.

The dorsal aspect of the milkmaid's forearm and hand displays the typical vesiculo-pustular lesions of cowpox. The largest pustule has begun to invaginate. In a few days the lesions will dry and crust over. Once the crusts drop off, the maid will no longer be able to transmit the virus to her close contacts.

racity of his claim that vaccination with his cowpox vaccine would protect against subsequent infection by smallpox. Jenner seemed pleased that the vaccine was "beginning to excite a general spirit of investigation." There were rumors about an entity called "spurious cowpox," which did not entirely protect vaccinees against smallpox. Was everything that was being advertised as Jenner's true product as pure as it ought to have been? I had been taught that Jenner used as his "vaccine" "lymph" extracted from farm workers whose pocks were believed to be a sure sign of cowpox. Was there a chance that some of the vaccinees were in fact inoculated with "deteriorated matter"? Could it be that the "matter," although possessing the authentic virus, might have "suffered a decomposition, either from putrefaction or from other causes . . ."; or from having "been taken from an ulcer in an advanced stage"; or from having produced sores "on the human skin from contact with some peculiar morbid matter generated by a horse"?

These eighteenth century smallpox vaccines, born and bred on the farms of England, were not subject to quality control. After Jenner produced his own lots of cowpox-derived vaccine, the manufacturing process caught on quickly, and farmers and the local physicians made and administered whatever they could. Sometimes the "matter" from the pocks of one vaccinee was used to vaccinate someone else, and so on and so on. Viruses are subject to deterioration and de-activation; sometimes they mutate, especially as they are passed from one subject to the next. Sometimes the vaccines get contaminated, particularly when they are inoculated with sharp objects, such as bird quills or shards of glass, as was routine in the late 1700s. Sometimes the infectious goop was allowed to dry in the open air or was corked inside a small vial. As Jenner noted, such a vaccine product was not likely "to possess all its specific properties at the end of three months."

I almost fell off my chair when on page 20 of his famous volume, Jenner mused:

> . . . whether the cowpox is a spontaneous disease in the cow, or is to be attributed to matter conveyed to the animal, as I have conceived, from the horse . . .
>
> I conceived this was its source from observing that where the cowpox had appeared among the dairies here (unless it could be traced to the introduction of an infected cow or servant) it had been preceded at the farm by a horse diseased in the manner already described, which horse had been attended by some of the milkers.

Jenner further observed that "morbid matter generated by the horse frequently communicates, in a casual way, a disease to the human subject so like the cowpox, that in many cases it would be difficult to make the distinction between one and the other."

I was satisfied to know now that the smallpox vaccine or vaccines, at least in their earliest European incarnations, were almost certainly random mixtures of at least three orthopoxviruses: smallpox, cowpox, and horsepox.

But I was surprised to learn that some of the live smallpox vaccines introduced subsequent to Jenner's product could be traced back to a soldier who succumbed to the disease during the Franco-Prussian War (1870–71) and whose viral strain ultimately made its way to Britain in 1907. Other vaccine stocks originated in England as early as the 1850s, and another came to Europe from Ecuador by way of Russia. The Tian Tan strain came from China.

In the absence of deep genomic analysis we will never know what passed for vaccinia or smallpox vaccine for nearly two hundred years between Jenner's time and the universal eradication of smallpox. I and those of my generation will never know what odd concoction of viral DNA made its way into our deltoid muscles during the vaccination process. We will never know whether snippets of DNA from whatever passed for cowpox or vaccinia may be sitting idly somewhere in our bodies. In retrospect, it was certainly more comforting to have been told that the stuff inoculated into the arms of hundreds of millions of people originated from a pustule on the otherwise perfect, alabaster hand of a diligent milkmaid rather than from the muck collecting on the hoof of a sturdy farm horse. And we will certainly never fully unravel the endless ways in which poxviruses carried in the wild by animals the world over have found ways of recombining before they accidentally make their way into humans who invade the habitats that serve as their homes or who consume their flesh when other sources of protein do not suffice.

One would hope that monkeypox could be eradicated as methodically and successfully as smallpox was about four decades ago, but recent reports from central Africa suggest otherwise. The Democratic Republic of Congo and Sudan appear to constitute the epicenter of the current twenty-first century epidemic. There is little argument with the fact that the discontinuation of universal smallpox vaccination, a practice that prevented infection with closely related poxviruses, was a critical precipitant of the ongoing African outbreaks, including those reported in parts of West Africa, as exemplified by the contaminated rodent shipment from Ghana that sparked the midwestern American outbreak in 2003.

It's not clear precisely when the current African epidemic began. The first case of monkeypox on the continent was confirmed in 1970, but it's likely that scattered cases had occurred earlier, only to be obscured by rare cases of smallpox that popped up during the final stages of the worldwide eradication program. The epidemiologists involved in the smallpox surveillance and control effort also discovered by chance that those persons who had been vaccinated against smallpox up to nineteen years previously had only a small chance of acquiring monkeypox when compared to the much higher infection rate in their unvaccinated neighbors. At the time of the most recent serosurvey in 2007–10, nearly half of all residents in Ghana and the Congo who were younger than twenty-six and who lived on the margins of a forest carried antibodies to monkeypox, meaning that they had been infected some time previously. Their infections were probably very mild, with few outward signs.

It has been suggested that those parts of equatorial Africa hardest hit are favored by a host of ecological characteristics conducive to monkeypox survival and spread. Among these are advantageous precipitation and temperature trends, deforestation, a beneficial altitude, and a large potential reservoir of susceptible primates and rodents. Human migration associated with political turmoil and economic deprivation almost certainly contributes to enhanced person-to-person transmission. Also, environmental and sociopolitical factors do not tell the entire story of the variable behaviors within the monkeypox lineage of viruses. Although the strength of antibody responses to different members of the virus family are similar, indicating the existence of sizable similarities in the structural proteins that constitute the virus, the pathogenic potential of the different viral cousins does vary, sometimes dramatically, despite similarities in their structure. The West African isolates, including the one that came to America, appear to have lost their talent for spreading readily from person to person, whereas the central African strains spread easily and with great ferocity. Some of the clinical characteristics that differentiate strains have been linked to subtle genetic differences, which become magnified and easier to quantify in the laboratory when related strains of monkeypox are used to infect susceptible animals, including prairie dogs, mice, and cynomolgus monkeys (*Macaca fascicularis*, a.k.a. crab-eating macaques). These measurable biologic effects include differences in mortality rates, in the intensity of end-organ pathology, and in the degree of multi-system spread. On a broader human scale, evidence to further support these clinical differences is based on the complete absence of fatalities among the monkeypox patients in the 2003 American outbreak, a finding of relative benignity that has rarely been reported from outbreaks in the Congo basin. Furthermore, although population studies have shown similar rates of antibody prevalence to monkeypox in West and Central Africa, ninety percent of all reported cases have emerged from Central, not West, Africa, pointing to a greater proclivity of the virus to spread and to cause more dramatic and more identifiable disease manifestations in the continent's center.

Because monkeypox is endemic in Central Africa and because this region is the most likely to contribute future cases that emigrate to the Western hemisphere, it's been a priority for epidemiologists to discover the full extent of the African disease and to define the routes the infection takes and the rate at which monkeypox spreads from index patients to first, second, and third generation contacts. It's also imperative to determine whether monkeypox can persist in the "wild" solely by person-to-person transmission, like measles or smallpox. The secondary attack rate for smallpox can be as high as 96% and for measles as high as 75%, even in the United States.

The *International Journal of Epidemiology* published a study examining these issues in 1988. Monkeypox surveillance was ramped up in Zaire beginning in 1980, soon after the termination of routine smallpox vaccination. Naturally, outbreaks were anticipated once the protective cover of smallpox vaccination was withdrawn. As expected, during the years 1980–84, soon after the termination of smallpox vaccination, two hundred and fifteen cases of monkeypox were reported by clinic and hospital staffs. Of this number, all came from sites within the dense tropical rain forest, and nearly all could be traced and studied in detail. These monkeypox victims were traced to one hundred and twenty-five discrete outbreaks, and two-thirds involved only a single patient. The disease detectives were surprised to find such a very low risk of person-to-person transmission. In fact, of the roughly two hundred reported cases, it was estimated that only about a quarter represented second, third, and fourth generation patients. It was even more surprising when heroic case-tracing efforts revealed that the one hundred and forty-seven primary and co-primary cases had a total of 1573 contacts, nearly eleven contacts for each primary or co-primary patient. Only a minority of these contacts got infected, and only a single chain of transmission in the Zaire epidemic persisted for five generations, the longest on record. Many of those who got infected were less than fifteen years of age and had never been vaccinated against smallpox. These young people who lacked the protection afforded by the smallpox vaccine comprised the most vulnerable group of susceptibles.

Scientists have identified five monkeypox genes that may be associated with differences in virulence. These researchers descend from a long line of geneticists who have been mesmerized by the possibility of finding clues to causal relationships between viral genes and the wide range of troubles that their coded proteins produce in the human hosts. History tells us, for example, that some strains of poliovirus merely caused self-limited episodes of summertime febrile illnesses, while others caused lifelong paralysis of the limbs or the diaphragm. Why? Why do some strains of Epstein-Barr virus, the etiologic agent of infectious mononucleosis, cause brief episodes of fever and lassitude, while others cause prolonged bouts of severe sore throat, swollen lymph glands, gigantic spleens, and weeks of unyielding fatigue? The answer is, we don't know. But if we could discover the genes responsible for these troublesome ills, we might discover a way some time in the future to fiddle with the genes, perhaps dampen their effects or silence them completely, thereby softening the blows meted out by these tiny viral control freaks.

Of course it's easy to point to better nutrition, superior medical care, and a mix of protective host immune factors in the American victims of monkeypox to account for their lower susceptibility and their better outcomes when compared to their less well-off African counterparts. But we always look first to the pathogens themselves to understand better what intrinsic viral factors support their troublemaking characteristics. Viral geneticists have focused on regions of the monkeypox viral genome that are known to maximize viral replication and primate-to-primate (including human) spread. Indeed, multiple nucleotide deletions and substitutions resulting in multiple gene sequence changes were observed when Zairean monkeypox viral DNA was compared with viral DNA extracted from West African isolates. Some of these genetic alterations affect proteins involved in the efficiency of viral replication, the extent of cell death that occurs in the wake of viral infection, and the enhanced severity of infection following inoculation of experimental monkeys by the intranasal or intracranial route. It has been hypothesized that many of these differences can be traced not solely to the virus but to heritable variations in the host immune response to infection.

Sometimes the immune response is so exuberant and poorly modulated that the host animal is harmed rather than helped. As an example, increased local tissue destruction may be mediated by complement, an essential immunologic actor in the cascade of the host response to infection. The inflammatory mediators produced in the course of these events often damage the host! But as usual, the virus and the host immune system perform a well-rehearsed dance in which a false step by one or the other, introduced by warring molecular choreographers, can change the intended performance for good or ill. Viruses are fierce invaders, and immunologists have identified several mechanisms by which the virus disarms one or another critical step in the cascade of beneficial immune responses. The route of infection, for example oral v. skin, and the time that has elapsed since prior smallpox vaccination or subclinical infection with a related or uncharacterized orthopoxvirus also influence the ultimate clinical outcome of monkeypox infection.

Just as it's critical to tease apart the factors that mediate the course of monkeypox infection in Africa, the only place on Earth where the disease regularly occurs, it's equally critical to tease apart the variables associated with spread of the West African viral variant in North America in order to prepare for the next unwelcome entry of whichever orthopox representative catches a ride on a contaminated human or rodent or monkey.

Human-to-human transmission of monkeypox was not documented in the 2003 Midwestern outbreak; all clinically apparent cases were traced to direct contact with a prairie dog. A case-control study was undertaken in which study subjects completed detailed questionnaires and had blood samples drawn for analysis of antibody levels to the responsible monkeypox isolate. Participants were asked whether they had previously received a smallpox vaccination. They were also asked whether they had recently engaged in activities like hunting, skinning, taxidermy, or animal trapping and if they had, or might have had, contact with animal feces, urine, saliva, or respiratory droplets and, lastly, whether they had been bitten or scratched. All the implicated prairie dogs had found their way into animal distributors' quarters, or private homes and veterinary clinics, where they were brought when they became ill. In the pet owners' homes, the prairie dogs shared the air with their human companions, were handled and petted, and had their cages wiped clean of uneaten food and excreta.

The thirty-three infected American patients fell into two categories: thirty had clinically apparent disease and met the strict criteria articulated in the investigatory case definition, and the remaining three did not meet the clinical case definition but did have elevated levels of fast-acting antibody. All case patients and uninfected controls had had exposure to ill or infected prairie dogs. Nine case-patients and fifteen controls had received more than one smallpox vaccination before it was abandoned in 1977. Among those with direct exposure to prairie dogs, the risk was greatest among those who had been scratched or who had cleaned the cages and touched the bedding. The risk was less, but not negligible, among those who had daily exposure to a sick animal or had touched the animals' rash or eye crusts. There was even some risk of infection in persons who had come within six feet of a sick prairie dog, but who never touched it. (This circumstance suggests some spread of infection by respiratory aerosols, as with influenza.) Remarkable confirmatory evidence of multiple routes of transmission of infection from the animals to humans was further amplified when dead prairie dogs were autopsied. Evidence of viral invasion was discovered in skin sores, ulcerated lesions on the tongue and conjunctivae and throughout the lungs. Experimental infection of prairie dogs also provided incontrovertible proof that respiratory, mucous membrane, cutaneous, and transdermal routes of transmission are possible. In this study, at least, the multiplicity of independent risk factors, potential routes of infection, and the varied composition of the study population made it impossible to determine whether prior smallpox vaccination provided a significantly decreased risk of acquiring monkeypox infection.

It may be impossible to apply the lessons learned in the United States to the natural history of monkeypox infection and its spread in equatorial Africa. First, the monkeypox clone that caused the outbreak in the US is now known to be more benign than the central African version and is much less likely to spread person to person. Second, the African infection is acquired during wild animal slaughters, preparation of the carcasses, and ingestion of the bushmeat. Also, African natives travel into the rainforest to hunt their quarry and have continuous exposure to animal nests and excreta in their natural environs. No one has studied the complex interplay of these varied factors vis-à-vis the chance of acquiring the African version of the infection and disease. To complicate matters, the immune status and nutritional state of susceptible Africans will also have to be factored into any calculus of risk. This is a formidable task and might only be undertaken should the central African endemic disease grow to epidemic proportions, cross international borders, and threaten members of other ethnic groups and cultures.

Despite the longstanding knowledge of the spread of monkeypox infection from animals to humans and, less often, from humans to humans in central Africa, some epidemiologists felt obliged to investigate the epidemiology of the slightly different monkeypox virus that circulates in Ghana. After all, that's where the American outbreak originated. What proportion of the mammals in the forests of Ghana were infected with monkeypox? How common were animal-to-human transmission events? What was the nature of these events; how intimate were the contacts and of what duration? It was hoped that the answers to these questions could be used to mitigate the effects of another illicit and foolhardy importation of a monkeypox-afflicted rat or monkey. Or, perhaps, another much-sought-after exotic pet infected with another poxvirus, perhaps buffalopox or horsepox or camelpox, or a poxvirus as yet undiscovered and unnamed.

A team of scientists comprising specialists from the Centers for Disease Control and Prevention, the Ghana Health Service, the Noguchi Memorial Institute for Medical Research in Accra, and the US Department of Agriculture National Wildlife Research Center in Fort Collins, Colorado, trekked to Ghana in the spring of 2004, about a year after the American outbreak of monkeypox. Their goal was to trap, examine, and study wild animals from the same area at the same time of year as the creatures that came to America in 2003. They also studied native Ghanaians who lived close to the implicated areas of rainforest, primarily to learn about those daily habits that might bring them into contact with animals in the forest and to determine how many might have already been infected with the monkeypox virus, albeit subclinically. Clinically apparent human monkeypox infection had never been described in Ghana.

The investigators focused especially on the three most abundant African rodent genera that were known to have been sent in the original shipment to the US, the giant pouched rat (*Cricetomys*), the African dormouse (*Graphiurus*, to be distinguished from the English dormouse; see Chapter 7 of *Alice in Wonderland*, "A Mad Tea-Party"), and the rope squirrel (*Funisciurus*). The squirrels and the pouched rats were collected from the forests northwest of Accra, and the dormice were collected from a site characterized by lowland scrub northeast of Accra. Altogether, two hundred and four animals were collected for study. Nearly two hundred had their blood drawn, but only two, one pouched rat and one rope squirrel, had antibodies at significant titers to monkeypox. A smattering of other rodents had antibodies, but in very low levels, suggesting that they had had milder infections, perhaps in the distant past. Also, only six percent of the animals from which tissue specimens were available had traces of the viral DNA.

A little more than fifty percent of the human residents of the area had detectable levels of antibody; more than half of them were older than twenty-three years of age. Most of these older natives had received smallpox vaccinations in early life, and the antibodies that were detected could have cross-reacted with the monkeypox and may well have been left over from earlier immunization with the closely related vaccinia virus, the putative agent of cowpox. Oddly, persons engaged in the exotic animal trade were not more likely to be antibody-positive than those who were not. In contrast, farmers were the group of natives most likely to have previously been exposed to monkeypox. They were often exposed to scavenging rodent pests and their nests, which are abundant in sylvan areas cleared for agriculture.

Somewhat to the scientists' disappointment, they identified no single rodent species as the predominant reservoir of monkeypox. Hence, in a quest to interrupt transmission to humans, there was no primary culprit to be followed, studied, and controlled. You could not warn unsuspecting farmers to focus attention on dormice or squirrels or big, fat rats with grubby overstuffed mouths. The affected animals might be found wherever forests grew and wherever forestland had been transformed to plots of farmland. The only practical option available was to warn the farmers to keep their distance from the various suspect animals during seasons when their populations are at peak levels. Animal density correlates closely with prevalence of infection for those pathogens that are transmitted efficiently from animal to animal and animal to human. In many parts of the Congo River basin and countries of West Africa, including Liberia and Sierra Leone, the risk of exposure of native rodents and monkeys to monkeypox virus is the rule rather than the exception.

I decided to focus most of this chapter on monkeypox for two reasons: First, there is evidence that this infection is now the most prevalent of the poxvirus diseases to which humans are susceptible. Second, monkeypox is the only one of this family of viruses to have caused a recent outbreak in the United States and to have set off a very necessary flurry of epidemiologic and virologic efforts. I admit that the story of contraband rodents from Africa meeting up with iconic, adorable American rodents and collectively bringing their plague to the heartland was too hard to resist. Of course, other closely related pox viruses have caused disease in humans, reminding us repeatedly that these incidents occur and that we continue to be at risk for wider spread of infection and outbreaks of serious illness should the right combination of viral genes and human susceptibility interact in the wake of the accidental or foolhardy introduction of sick animals into an environment where Mother Nature never intended them to be.

In researching this chapter, I located a report of tanapox infection in a Dartmouth College student who developed a febrile illness characterized by severe headache, backache, and a tender, papular eruption on her elbow and calf. The lesions appeared both during and after a sojourn in the Republic of Congo, where the student spent eight weeks caring for orphaned chimpanzees. Tanapox infection is endemic in equatorial Africa, where the disease, always accompanied by a rash, is not unusual. But only four cases had been reported in the United States prior to 2015. Swollen and tender lymph nodes in the armpits and around the elbow appeared in spots that drained the affected areas of skin. Quite reasonably, the diagnosis of the student's pox eluded her first American caregivers. They wondered whether she might have tropical ulcer or some sort of atypical mycobacterial infection (a relative of cutaneous tuberculosis) or, as unlikely as it might seem, cutaneous anthrax. As soon as skin biopsy samples became available, the pathologists, electron microscopists, bacteriologists, and molecular virologists got to work. The tissue stains and the electron microscope images quickly revealed the hallmarks of a poxvirus etiology. Polymerase chain reaction studies demonstrated the presence of tanapox DNA sequences, and definitively ruled out the possibility of cowpox, monkeypox, vaccinia virus, and variola (smallpox) virus, four other species well-known to cause infections in humans. Prompt and accurate diagnosis put a stop to predictable panic and saved lots of money by forestalling lengthy public health investigations and interventions.

Tanapox virus was first isolated from schoolchildren during a 1957 outbreak in a small village in Kenya, which was near the Tana River. Following a second outbreak in 1962 in another part of the same river valley, it was learned that tanapox is enzootic and finds its reservoir in nonhuman primates. Although insects may be responsible for spread of tanapox among monkeys, transmission from primates to humans and from humans to humans is believed to be direct.

The college student recovered uneventfully in the absence of any effective antiviral chemotherapy. The lesions reached a maximum size of about two centimeters in diameter. They became umbilicated, developed necrotic crusts, ulcerated, and then deepened to form nodules. They were completely gone in six weeks, and the only hint of their former existence was small scars.

Cowpox has not been forgotten, even though it virtually never makes it onto the list of differentials which doctors construct to diagnose patients with pustular or vesicular skin lesions. On that list of blistery diseases, herpes simplex, yes; varicella-zoster, yes; an odd spider bite reaction, perhaps; but not smallpox or monkeypox or cowpox. So we cannot fault the doctors who questioned and examined and puzzled over an unfortunate seven-year-old Swedish girl who got infected with cowpox in the year 2000.

This poor girl came to her doctor complaining of an extensive reddish swelling of the skin and tissues below her right ear. Her condition worsened rapidly; the entire right side of her face and neck became edematous just before the swelling spread to the left side and both eyelids. Local lymph nodes became grossly swollen and painful, and a necrotic mass erupted inside her nose. Hyperbaric oxygen therapy was introduced in order to combat the anaerobic bacteria that were presumed to be replicating in the abscess cavities within the ever-enlarging lymph nodes. These abscesses were ultimately incised. Rubber drains were placed to hasten the flow of pus. The patient ultimately recovered, and only small scars remained as evidence of her critical ordeal.

I tell this story to document the possibility that poxvirus infections may appear without warning even in the wealthy industrialized countries of the world. Whence do they originate and how do they get from point A to point B? Identifying the simultaneous variables that encourage their spread is a matter of endless research. Did the Swedish girl's recent trip to Spain have anything to do with her subsequent infection? We'll never know. The girl reported that she had two cats which she was very fond of. One of the cats often licked her face and neck, and as an act of great affection its tongue sometimes entered her nostrils. The cats became prime suspects, but they were healthy; they had no sores on their paws or mouths, and when their blood was tested for antibodies to cowpox at the Veterinary Institute in Uppsala, the results were negative. Did something go wrong with the testing procedure? Were the cats infected with cowpox so recently that there was insufficient time for them to mount a detectable antibody response? Hypotheses ran rampant, but the mystery was solved when goop from the little girl's festering nasal sore revealed cowpox DNA upon PCR analysis, which was undertaken at the CDC in Atlanta, Georgia. Beginning in the early 1990s, multiple reports in the medical literature have proven that dogs and cats, especially hunting cats, sometimes shed cowpox virus in their oral secretions. How did the virus get there? By way of rats, voles, and foxes! Feline carnivores get infected when they take as their prey members of those feral species that are known to be the reservoir for cowpox. In one European study of the prevalence of antibody to cowpox in various species, domestic cats were found to be the most common hosts for the virus. Cats are the source of infection in most human cases, and girls younger than twelve are those most frequently reported.

We know that household members and healthcare workers are first in line as potential victims of infectious diseases that are brought into their midst. But we sometimes forget that laboratory workers are also at risk, particularly bacteriologists, virologists, and pathologists who day in and day out handle human blood, urine, feces, sputum, and tissue from all parts of the body. They are, in fact, among the "first responders" whose job it is to give a name to an intrusive and lethal pathogen. There is an annual International Meeting of Emerging Diseases and Surveillance and periodic conferences of the CDC's Epidemic Intelligence Service. During some such assemblies the following story of occupationally-acquired cowpox was told: A researcher who worked with a non-orthopox virus, NOC or Chordopoxvirus, discovered a tender, ulcerated lesion on a finger. She also complained of fever, body aches, headache, and lymphadenopathy. After seeing multiple physicians, she agreed to have samples of the sore's exudates sent for specialized virologic testing. Cowpox virus DNA was ultimately found in the specimen; but by that time the sore had already healed. Therefore, public health officials first got involved only after the illness had passed. Cowpox is not endemic in the United States, so the investigators were charged with determining whether the infection was acquired somewhere other than the lab (a scary possibility), or as a result of some accidental exposure in the lab. It was learned that stocks of cowpox virus were, indeed, stored in deep freeze in the patient's workplace, but cowpox was not currently being used in ongoing experiments. In fact, no one had been working with cowpox for the previous five years. Also, in contrast to previous years, the laboratory was no longer requiring its workers to be immunized prophylactically with vaccinia virus (smallpox vaccine) because the lab was no longer working with any poxviruses that were pathogenic for humans.

Virology laboratories are charged with maintaining the most excruciatingly stringent safety behaviors. Many operations are performed using sterile technique, not unlike what we're used to seeing in the operating room. The routines are bi-directional: we don't want the lab workers to get infected with the pathogens they're working with, and we don't want the virus stocks or cell culture stocks to become contaminated with any members of the colossal circus of bacteria and fungi that comprise our own personal microbiomes. More than one virology lab has had to close its doors temporarily when specimens or virus stocks or cell culture stocks became sullied with the mycoplasmas or staphylococci or rhinoviruses or yeasts that meandered silently from hand or mouth or nose or eye into pristine and nutritious sterile culture medium or a flask of gently floating or sessile mammalian cells.

The lengthy investigations of the researcher's lab uncovered a number of breaks in technique, some as yet unshared but critical information, and some unwise storage habits. It turns out that the NOC virus stocks, which were used by all members of the laboratory staff, shared space in the same rack of reagents in the same freezer as the cowpox virus stocks. I can attest, as a result of years of work in a virology research lab, that labels on decades-old tubes sometimes wear out or become illegible, if only momentarily, by dense accumulations of frost. (Viral stocks are usually kept at −70° centigrade.)

Every time you open the ultralow temperature freezer, some room air, replete with its water vapor, gets inside. The vapor turns to frost almost immediately.

It turned out that at some point a few of the experimental mice which had been inoculated by the research tech developed "umbilicated bumps on their skin with a central area of necrosis." These lesions were attributed to a staph infection. Not true! A review of notebooks revealed that the mice were injected with fluid from a NOC virus stock that was later found to be contaminated with cowpox virus. Also, other mice were injected with pure cowpox virus, drawn from tubes that were simply misidentified as they emerged from the freezer. When interviewed, the patient recalled that gloves were used inconsistently, in violation of protocol, when handling live virus or cell cultures, particularly when lab workers moved items from the ultra-cold freezer to the warm incubator. After multiple virus and cell stocks and environmental swab samples were analyzed exhaustively by the most precise molecular techniques, it was found that four NOC stock vials were also positive for cowpox DNA. One of these vials was the same one used by the researcher prior to her illness and, as it turns out, it was the same vial from which virus was extracted and used to inoculate the mice who later developed the suspect lesions. Three of twenty environmental swabs were also positive for orthopoxvirus DNA: pipettors kept next to the biological safety cabinet; the handle of the freezer; and the exterior of the freezer box of NOC stock viruses.

Concern was aroused when it was later discovered that the cowpox viral DNA sampled directly from the patient's sore contained recombinant sequences, that is, bits of DNA from a virus other than NOC. I reveal this iota of data to state a simple fact, "things happen!" Humans are fallible. Laboratories are busy places replete with much equipment and many busy people, often performing multiple tasks. Sometimes people and machines fail to behave optimally. In the paper that reported this incident and in subsequent investigations, scientists made these recommendations: a) that there be immediate notification to health authorities of any suspect or confirmed orthopoxvirus infection; b) occupational exposure to an unusual pathogen should be anticipated when laboratory workers present with unexplained illnesses; c) when zoonotic diseases appear that are not endemic to a particular geographic area, prompt testing and forthright reporting to institutional and public health authorities should be mandated in order to prevent spread of the infection and immediately put in place reparative measures. Finally, it was suggested that vaccination with the smallpox/vaccinia virus vaccine be required of all individuals who might be at risk for exposure to any orthopoxvirus, monkeypox or cowpox, "in the wild" or in the lab. In this particular instance, trepidation arose due to the real possibility that a novel, recombinant virus, with unknown potential for person-to-person spread and virulence, might make its way beyond the confines of the laboratory.

CHAPTER 5

A Civet? What's a Civet?

SARS (SEVERE ACUTE RESPIRATORY SYNDROME)

> At Dongmen market, almost everything that's vaguely edible is for sale . . . Geese, ducks, chickens, pigeons, doves, and wild birds are packed wing-to-wing in metal cages stacked two and three high, their minders napping on top . . . Nearby, rabbits are squashed in cages, and turtles and crabs in huge metal tubs scramble over each other . . .
>
> But the market has lost some of its legendary variety. Masked palm civets, for one, are missing . . . These distinctive cat-like creatures were readily available at Dongmen and similar markets across the province. Restaurateurs bought them for meat, said to be tasty and fabled to strengthen the body against winter chills.
>
> *Science Magazine*, July 18, 2003

On February 11, 2003, the Chinese Ministry of Health notified the World Health Organization of the sudden appearance of a severe acute respiratory syndrome (soon dubbed SARS), a disease of unknown etiology. It was spreading in Guangdong province in southern China. The disease had been out of control since November 2002. A large number of its earliest victims were healthcare workers or household contacts of stricken victims who had, in turn, been exposed to persons recently returned from the south. One middle-aged man, who returned to his home in Vietnam after a trip to China, infected hospital workers who cared for him in Hanoi. A similar case cluster suddenly broke loose in a Hong Kong hospital. Soon localized outbreaks erupted throughout Southeast Asia. Then a man who had recently traveled to China fell acutely ill in Toronto, Canada.

Like so many similar respiratory illnesses, the disease first produces a high fever, muscle aches, chills, rigors, and sore throat. Some patients developed shortness of breath and then pneumonia. The most unfortunate ones needed hospitalization and ultimately required mechanical ventilation. Then white blood cell and platelet counts fell. SARS appeared to be very contagious; the attack rate among the first contacts of patients was about fifty percent. Many of these victims were hospital workers.

The CDC in Atlanta, Georgia, issued a travel advisory in March, 2003 and ordered enhanced domestic surveillance for SARS. Travelers were advised to cancel all nonessential trips to Hong Kong, Guangdong province, and Hanoi. Airline passengers returning from the three high-risk areas were advised to contact their doctors if they developed fevers and respiratory problems. By mid-March, eleven people were under the CDC's surveillance with suspected cases of SARS. Their chest X-rays showed signs of pneumonia. Doctors throughout the United States were advised to measure their patients' blood oxygen levels, to get chest X-rays, and to submit sputum samples to the lab to test for known viral pathogens, such as influenza and respiratory syncytial virus, and for common bacterial pathogens. Lab techs were urged to save aliquots of blood, biopsy specimens, and sputa, all of which were to be mailed under special conditions to the CDC or to state health departments. Most important, because the exact mode of transmission of SARS was unknown, clinicians caring for suspected SARS patients were instructed to use standard hand hygiene along with respiratory (paper masks or preferably heavy duty N-95 respirators) and contact (gowns and gloves) precautions. Some hospitals added eye goggles to the mix, because viruses may find the eye a convenient portal of entry.

In the era of globalization, with its vast numbers of international travelers, managing a worldwide disease outbreak, particularly one spread by the respiratory route like influenza, is unimaginably tough. It's been estimated that in the first years of the twenty-first century about half a million residents of China, Hong Kong, and Vietnam traveled to the US annually. Approximately that same number of people are hospitalized with pneumonia in American hospitals each year. And in at least half of these cases the etiologic agent is never discovered. So how, in this huge mix of febrile, coughing, gasping patients, do you pick out the few with SARS when you have not yet figured out what causes SARS and how to treat it? In April, 2003, the Centers for Disease Control opened an Emergency Operations Center and asked local health departments and international airlines, cruise ships, and cargo carriers to notify CDC of suspected cases, of which there would likely be thousands.

In the eight months following the first announcement of the SARS epidemic, approximately eight thousand probable cases from twenty-nine countries were reported to the World Health Organization. Twenty-nine cases were reported from the US. About ten percent of the first SARS patients who met the official case definition had died; none were US citizens. We can't control epidemics until we know how and where they start. Once we do know, it might be possible to find a way to stop the epidemic dead in its tracks, to smother the sparks that may soon erupt into flames. The spread of SARS had features that resembled earlier influenza epidemics. They almost always emerged from the Far East, primarily China, and the earliest victims were likely pig and poultry farmers or food handlers. In each instance, disease spread rapidly among co-workers and household members before wending its way among villagers, townspeople, city dwellers, and, lastly, inhabitants of other countries who were stricken while far from home or on the planes and ships that carried them back.

Meirion Evans and epidemiologists from Cardiff, Wales, were among the first microbe hunters determined to find the animal reservoir of the SARS microorganism and, if possible, to neutralize it on its home ground. Having reviewed raw data provided to them by a host of Chinese scientists, they soon determined that most of the earliest SARS victims worked in the animal and food industries and often lived close to live animal markets. Evans also learned that foodies in Guangdong were especially fond of exotic meats. The meats were obtained from wild animals that were caged on the restaurant premises and slaughtered there so as to be served most efficiently to the waiting diners. Early in the investigation, they ruled out pigs, chickens, geese, and ducks as suspects. Had these common food proteins been responsible, many more farmers, chefs, and their contacts would have fallen victim at the outbreak's beginning.

Teams of epidemiologists from all over the world descended upon southern China. The first to correctly identify the infected animal that triggered this international disaster would garner instant glory. Winning renown as the brightest and best of the world's medical sleuthing teams is the blue ribbon, the Olympic gold medal of epidemiology. The competition began when mostly staid, focused, selfless, and obsessed scientists, usually sequestered in their labs, broke their tethers and joined the race to the finish line. In April, 2003, the SARS coronavirus was provisionally identified, so the epidemiologists had the clear-cut job of collecting blood samples from a slew of animals in the Guandong market and its environs. They would test for antibodies to the provisional virus. This would reveal which animals were currently or had previously been infected. A team from the University of Hong Kong and the Shenzhen Center for Disease Control and Prevention determined that two animal species harbored the SARS virus, masked civets and raccoon dogs. A single Chinese ferret badger also had antibodies to SARS circulating in its blood. It seemed an easy step now to nip the disease in the bud and interrupt the first stage in the epidemic cascade. Civets, raccoon dogs, and ferret badgers were removed from the markets, and their flesh was removed from the menus of nearby restaurants. But it wasn't that simple.

Another team from the China Agriculture University working in a slightly different location was unable to find any evidence of the SARS virus in civets or in any other species they investigated. What accounted for this discrepancy? Were there localized spheres of infection, discrete geographic niches in which only the perfect array of animals, plants, and humans could provide a good home for the virus? The WHO was commissioned to find out; it gathered teams from all over the world. The international team of scientists and a few others from the CDC in Shenzhen, China, traveled to the Dongmen animal market, less than a half hour from the border with Hong Kong. There they literally borrowed from their vendors twenty-five animals (which they promised to return), representing eight exotic species. Wearing impermeable body protection gear, the scientists collected blood and nasal and rectal swabs. No one knew where this new virus liked to build its home and multiply. Representatives of three of the eight species examined carried the SARS virus or antibodies to the virus, the same three species of exotic mammals that had already been implicated. The remaining five species were negative for SARS. Not surprisingly, antibod-

Civet (*Civettictis civetta*). Photograph by Kajornyot Wildlife Photography/ Shutterstock.com.

ies to SARS were also found in some of the healthy market workers, but no such antibodies were found among members of the wider population.

These reports did not satisfy competing teams from other regions in China. These competitors wanted to be the victors in the battle of the "firsts," but they delayed publicizing their arguments until the competing groups had presented their findings at scientific meetings or in well-regarded peer-reviewed journals. Ultimately, in an honest attempt at some sort of concord, the experts agreed that civets "played a role" in the transmission cycle of SARS, but that much work still needed to be done to trace the virus back to its ultimate source. Perhaps the civets picked up the virus following capture, particularly in the teeming, dirty animal markets of southern China.

In other countries, investigators planned to experiment more fully with the civet isolate of SARS. They would inoculate monkeys with the virus to learn about the susceptibility of primates to infection. In the meantime, the dissenting agricultural university group proceeded to gather more wild animal representatives from the Beijing area and an additional fifty-four species from six other provinces. The researchers also included some domestic animals in their search and added a few palm civets from Guangdong, where greedy sellers were still purveying their prize animals to restaurateurs despite the country-wide ban! The agricultural university group found not a single animal infected with SARS, but did find a genetically related but not identical virus in one civet.

Accusations rumbled back and forth. Some groups griped about the failure of other groups to share all their data, thus foiling their chances of refuting specific claims of certitude. Others argued that by branding civets as the principal perpetrators of the growing outbreak, the report would have deleterious effects on professional experts in civet husbandry and capture, on civet salesmen, and on savvy chefs and restaurateurs. There would also be disappointed gourmets whose taste for civet flesh was big business. One group of civet farmers in Taiwan threatened to sue the University of Hong Kong for ruining their business. In light of all the controversy and threats of legal action, the scientists at the Shenzhen CDC made clear that they had never said civets were the natural reservoir of the SARS virus. Perhaps civets were merely one link in a chain of viral transmission. Perhaps they had been infected in the wild by some other animals, maybe bats.

The magazine *Science* queried epidemiologists on site, and some referred to a 1998 Malaysian outbreak of Nipah virus (see chapter 9) infection in which fruit bats, the natural reservoir, transmitted the virus to pigs, who then spread the disease to humans. So perhaps bats originated the SARS epidemic in southern China. It's quite common for the findings of competent scientists to conflict with one another. They may use different methodologies, or they may begin their inquiry with different hypotheses, starting points that evolve into disparate approaches to a question. Hence, there was no expression of surprise or criticism when the results from epidemiologic explorations in one part of China differed from findings in other parts of China. One question remained: Were civets and raccoon dogs the sole progenitors of the SARS outbreak or were they not? Would their numbers have to be reduced severely in order to quell the epidemic? Could a way be found to reduce the spread of SARS-CoV, SARS coronavirus, in its natural habitat? Perhaps there was another animal family altogether that had long before been the ultimate reservoir for the virus. Did this animal reservoir transmit SARS to other secondary mammalian species that had contact with one another by virtue of sharing the same ecosystem?

One fact was certain—humans were not the reservoir! SARS-CoV was brand new, unrelated to the other coronavirus families that were known for decades to cause respiratory and gastrointestinal disease in humans, almost always of a mild nature. SARS-CoV also differed genetically from the numerous coronaviruses of domestic animals, including poultry and pigs.

In 2004, bats rose quickly to the top of the list of suspects. Bats have long-been known to be carriers of myriad viruses, some, like rabies, are pathogenic for humans. Their wide geographic range, their habit of nesting in tight-knit groups, and, for some, their propensity to bite, make them perfect vehicles of virus transfer and dissemination. In addition, the viruses bats carry only sometimes cause symptoms or disease in their hosts. This symbiotic relationship allows the viruses to replicate quite happily, sometimes for years, without killing the hosts that provide them with unrequited domestic tranquility.

Scientists from the Wuhan Institute of Virology and Duke-NUS, using genetic sequencing, discovered the presence of SARS-like CoV in bats of the genus Rhinolophus, horseshoe bats. Four horseshoe bat species from different regions of the

People's Republic had evidence of infection based on the results of PCR and serologic tests. Some of the affected bats populated regions as far apart as 1,000 kilometers. In some regions of China, more than eighty percent of bats had antibody to a major protein of SARS-CoV. Widespread geographic range and high antibody prevalence characterize reservoir species of mammals. The genetic sequences of SARS-CoVs isolated from bats were far more diverse than the SARS-CoVs of humans and civets, suggesting that the long association of SARS-CoV and bats allowed greater opportunity for the bat viruses to mutate into a mix of closely related viral species, one of which found a welcome haven in humans. In fact, only about ten percent of the entire genetic sequence varies among most bat SARS viruses. This same degree of sequence identity is seen in some bat SARS viruses and the SARS-CoVs of humans and civets.

These SARS viruses are as closely related to one another as kissing cousins or as branches on a tree. The tree-branch analogy is pertinent because geneticists utilize that imagery when constructing their phylogenetic trees or dendrograms. These visual constructs display as branches on a tree the relationships between the amino acid sequences of important viral proteins. The relationships are deduced from detailed analysis of the genetic sequences that encode the proteins. The bigger branches that spring directly from the tree trunk are less similar than the secondary and tertiary branches. The little branchlets, far from the trunk but very close to one another, represent those viral sequences that are further along the evolutionary tree and hence more closely related genetically. It's now known that bat coronaviruses tend to be species-specific, that is, each bat species within a particular geographic range carries its own SARS virus, and these bat-coronavirus duos stay together as the bats travel from place to place. In the case of SARS, this close virus-bat connection is not good news for those people who live close to the SARS-infected bats, as they fly from place to place and bite scampering ground animals or contaminate them with infected feces. There is one caveat—what we know of the epidemiology of SARS indicates that mammalian interlopers, in this case civets or raccoon dogs, must bear the burden of the virus before it travels in sequence, to the animal handler, the food handler, and the dinner table. This transmission cycle is likely to occur only in those unique locales, such as southern China, where huge live animal markets prosper and where prosperous Chinese gourmands relish civet flesh. The die is cast for the rest of us when one of these diners, newly infected with the virus, and carrying and shedding it, meets up with others who are merely going about their daily activities or staying at a hotel where an infected person sneezes in an elevator or touches a door handle.

Humans were already counted among the more socially advanced victims of SARS, but exactly how does the virus do its dirty work in primates? What's the most efficient route of infection? How widely does the virus spread in the body? How does the body respond immunologically? What is the mortality rate? Are there possible clues to an effective treatment, or perhaps a cure?

SARS is caused by a coronavirus. On April 12, 2003, exactly two months after the first SARS outbreak was reported by the Chinese to the WHO, the virus was isolated almost simultaneously in laboratories in New York, San Francisco, Manila, Hong Kong, and Toronto. The sequencing of the entire genome of the virus was accomplished by collaborating research teams at the British Columbia Centre for Disease Control and the National Microbiology Laboratory in Winnipeg, Manitoba. The isolates were collected from patients in Toronto, the epicenter of the North American outbreak.

Coronaviruses carry a single strand of RNA as their genetic code. The RNA strand comprises thirteen genes which encode fourteen proteins. Like other coronaviruses, such as the one which causes Middle East Respiratory Syndrome, or MERS (see chapter 1), the SARS genome encodes the proteins which together create the physical structure of the virus: a nucleocapsid, the protein which encases the RNA; a membrane protein with its characteristic spikes that form the corona, or crown, at the viral surface; and a lipid-containing envelope. There are other accessory proteins, some of which orchestrate the virus's replicative machinery, and others whose functions remain unknown. The SARS virus is closely related to two other coronavirus species, but is much different genetically from another three groups of coronaviruses. It's believed that the SARS coronavirus has co-evolved with its animal hosts, possibly bats and some ground foragers, like civets and badgers, for eons before achieving the capacity to jump successfully to primates, including humans.

There is now strong evidence to support this belief; SARS virus isolates derived directly from masked palm civets contain twenty-nine more nucleotide "beads" in their RNA strand, when compared with SARS RNAs isolated from human SARS patients. As reported in the magazine *Science*, coronavirologist Peter Rottier of Utrecht University believes that the shorter stretch of RNA may have made the virus more adept at first finding refuge in humans. Or perhaps some of the earliest SARS coronaviruses derived from humans lost the twenty-nine nucleotides only after passing through a successive number of human hosts. Evidence to support this theory comes from analysis of the virus collected from one of the earliest victims of the Guangdong outbreak. His SARS virus did carry those additional twenty-nine nucleotides. Viruses are always changing their stripes, especially as they pass through successive hosts in different species. It's a brilliant trait particularly common to RNA viruses. The goal is to infect as many victims as possible, to find safe haven in a variety of species, and to mutate as efficiently as possible, with the ultimate goal of successful reproduction and efficient spread. The ultimate "plan" is to generate more rapacious baby viruses that are imbued with the guile to identify and attack as many unprepared cellular hosts as possible in the least amount of time.

Before the Age of Exploration reached their shores, inhabitants of the Western hemisphere were safe from sporadic threats, both bellicose and biologic, posed by events erupting far across the oceans to the east and west. Then the Europeans came in search of gold, bringing conquest and forced eviction from native lands. Along with avarice in their hearts, they brought microbes in their blood and in festering sores on their skin. Slaughter by the sword was accompanied by somewhat slower deaths caused by smallpox, measles, and syphilis. Neither Aztec, Inca, nor Mayan empires of Central and South America nor the indigenous tribes in the north had previously encountered these diseases, so communal or "herd immunity" had not been created in the local populations. Without natural defenses, the natives of the Western hemisphere died in startling numbers.

Because modern transportation has drastically increased human contact between the two hemispheres, the microbial threats are in some ways more immediate. Human beings continue to be the vehicles of doom. When SARS broke out in the Far East, those in the West might have felt temporarily safe. Then a traveler, recently returned from China, arrived in Toronto and precipitated an outbreak. Fear of a pandemic soon gripped well-informed citizens and government officials throughout the world.

On February 23, 2003, a married couple returned to Canada from visiting relatives in Hong Kong. While there, the husband and wife stayed in a hotel. Just a few days prior to their arrival, another hotel guest, later identified as the index case for the entire East-West outbreak, occupied a room on the same hotel floor. The husband and wife only stayed in the hotel at night; they spent entire days at their son's place. They came home to the apartment in Toronto that they shared with two sons, a daughter-in-law, and their infant grandson. The old lady, with a history of coronary artery disease and diabetes, was the first to succumb. Two days after her return, she developed a fever, a sore throat, and a mild but persistent cough. She went to her doctor. Other than a red throat, her exam was normal. The doctor prescribed a course of antibiotics anyway. She was sent home, her cough got worse, and she became dyspneic. She died a mere nine days into her illness. The woman's son, a forty-three-year-old with a history of hypertension and type 2 diabetes, fell ill two days after the start of his mother's illness. In rapid succession, he developed a fever, sweats, a cough, chest pain, and shortness of breath. A chest X-ray showed consolidation in two lobes of the right lung. He was given an antibiotic. His symptoms failed to respond, and he returned to the hospital, with a fever of 104° F and a greatly diminished blood oxygen saturation level. His chest X-ray indicated that the disease had spread to his left lung. He was treated with broad spectrum antibiotics, including those for tuberculosis, which could not adequately be ruled out. His rapid deterioration prompted the need for mechanical ventilation and intensive support for multiple failing organs. He died six days after admission. A post-mortem exam revealed diffuse, bilateral pulmonary damage and a suffusion of the interstitial tissues of the lung with white blood cells of various lineages, a finding most compatible with overwhelming viral infection. A search for the most likely culprits was unrevealing. Foci of necrosis and hemorrhage were also found throughout the liver and spleen.

Concerns about a possible family-wide outbreak of tuberculosis continued to nag at the minds of their doctors. All the remaining adults and children were either febrile or coughing or both. The adults had abnormal chest films, but the exposed children had no visible lung pathology. At the time of this case cluster, doctors in Canada did not yet have definitive diagnostic tests for SARS. However, eight family members met the official case definitions for either "suspected SARS" or "probable SARS" as set forth by official public health authorities in different parts of the world. The sickest four in the family were admitted to the hospital, three of them to intensive care units; all received wide-spectrum antibiotics and two antivirals, none of which was specific for the treatment of the SARS coronavirus. The four recovered, but some suffered residual breathing difficulties for up to a month following discharge.

Three additional patients appeared on the scene just as the last member of the Chinese family received care. One of the three was a previously healthy nurse who had cared for a sick man and wife in the hospital outpatient clinic; another was an elderly man of non-Asian descent who came to the hospital emergency room with a bout of atrial fibrillation. During this stay in the ER he was separated from the gurney of one of the aforementioned SARS patients by a cotton curtain. The two shared this space overnight while the SARS patient waited for an inpatient bed to be available. The old man developed respiratory symptoms two days after his arrhythmia had been successfully treated and he was sent home. He returned with high fever, dyspnea, nonproductive cough, hypothermia, and oxygen desaturation. Despite intensive care, artificial ventilation, and vigorous anti-infective treatment, he died twelve days after illness onset.

Travelers from all parts of North America visit all parts of the Far East, so by chance alone one would have expected to see other SARS suspects popping up in places far from Toronto. Sure enough, a couple returned to Vancouver from a trip to Hong Kong on March 6, 2003. The couple had stayed in the same hotel as the Chinese man and wife in the Toronto case, but not on the same floor. They did not eat in the hotel's restaurants and did not visit with any of the other guests. The husband got sick during a brief stopover in Bali on his way back to Canada. He was admitted to an intensive care unit in Indonesia with a temperature of 102°F and a blood oxygen saturation half the normal value. He required intubation and mechanical ventilation.

Detailed review of these early cases revealed themes and clues to SARS' pathogenesis that set the foundation for more targeted diagnosis and treatment of all the multinational victims who soon fell ill. First, SARS appeared to be highly contagious; merely sharing space in a multistory hotel or breathing the same air in an emergency room could put one at risk. Second, SARS could spread to far-flung parts of the world in a matter of weeks; international travel made this very possible. Third, there might be genetic and immunologic risks to one's susceptibility. It is true that the disease started in the Far East, but did this point to a particular genetic predisposition among Asians or merely a cultural one, such as a tendency to inhabit spaces close to the wild mammals which would soon appear on your dinner table because of your fervent belief that a civet's or a badger's flesh would make you stronger or healthier or sexier? Certainly, bad luck also played a role, as it does in so many other instances of misfortune—merely being in the wrong place at the wrong time and bringing to that place and time a history of diabetes or hypertension or smoking, or some other all too common underlying condition that enhances the virus's idiosyncratic method of attack.

All in all, twelve guests who stayed in the now infamous Hong Kong hotel became infected with the SARS coronavirus; seven had occupied rooms on the ninth floor, where the first two patients had stayed. These twelve guests had carried the pathogen to their home countries—Vietnam, Singapore, Canada, Ireland, and the United States. By mid-April, nearly 3,400 cases of SARS and 165 deaths had been reported from twenty-seven countries.

All deaths matter, but one in particular brought international attention to the risks assumed by microbe hunters who travel to contaminated places and get down and dirty as they seek to discover as much as they can as fast as they can about the pathogens that occasionally shake humans out of our customary torpor.

Dr. Carlo Urbani, an infectious disease expert from the World Health Organization, who had been president of Médecins Sans Frontières, Italy, left his parasitology lab and agreed to investigate an outbreak of a particularly virulent form of pneumonia, as yet unnamed, at the Vietnam French Hospital. As usual, a new strain of avian influenza was on the first list of "rule-outs." He collected clinical samples for analysis, but more importantly reinforced already strict isolation and infection control procedures. Under his guidance, an isolation ward was created and kept guarded. More than half the patients were healthcare workers, some of whom quarantined themselves at the hospital, thereby putting themselves at even greater risk, rather than return to friends and family in the community. Dr. Urbani was alarmed at the fast pace of the disease's spread and urged the WHO to arrange an emergency meeting on March 9, 2003, with Vietnam's Vice Minister of Health. After the meeting, the entire Vietnam French Hospital was quarantined, and newly-stricken patients were housed in other freshly prepared hospitals that were soon equipped with infection control suits and kits supplied by volunteers from the WHO, the Centers for Disease Control and Prevention, and Médecins Sans Frontières.

These measures, taken in the open, in view of the entire world, ended the outbreak in Vietnam. This success, unfortunately, was not to be witnessed by Dr. Urbani. On a flight to Bangkok he got sick, and upon his arrival a colleague called an ambulance. For the next eighteen days, he fought a progressively downhill battle with SARS; he died on March 29. In special recognition of his sacrifice and his lifelong struggle against other diseases, such as AIDS, tuberculosis, and malaria, the *New England Journal of Medicine* published a testimonial praising his "selfless devotion to medicine."

Already distinguished as part of the Doctors Without Borders team that received the Nobel Peace Prize in 1999, Dr. Urbani was one among a host of healthcare workers who died in the line of work. In Hong Kong, about a quarter of all patients with SARS were doctors and nurses. Among these was the chief executive of the island nation's hospital authority. The high death rate was partly attributed to the ever-waning stocks of protective gear and other infection control equipment. There were rumors that the manufacturers of these materials withheld their products from export to better provide for their own countries should SARS arrive on their shores.

The international effort to understand the epidemiology and clinical manifestations of SARS and to devise effective infection control modalities was bolstered by an equally worthy worldwide effort to fully study the SARS coronavirus: to discover the intricacies of its behavior in cell culture and laboratory animals, to delve deeply into its structure and genetic composition, and to develop theories as to possible ways to intervene pharmacologically or, perhaps, by means of vaccination.

The earliest steps in virus identification are well-known and replete with historical precedent. Specimens from the patient are usually taken from anatomical sites proximate to the presenting symptoms—from sputum or nasopharyngeal secretions in the case of influenza or pneumonia and from stool in the case of gastrointestinal upsets such as vomiting and diarrhea. These specimens are inoculated into living tissue cultures, often single sheets of cells, sometimes of human origin, sometimes of other animal origin. The cells adhere to the surface of plastic or glass flasks that are bathed in enriched medium designed to supply all the nourishment the cells need. This complete diet supports them through the arduous process of giving birth to tens of thousands or millions of viral progeny. When the tissue culture cell sheets are in the throes of viral replication, they display pathologic changes that are characteristic of each viral family. Herpes simplex virus, the cold sore virus, causes the cells to swell dramatically, producing an effect called "ballooning degeneration." Cytomegalovirus, as its name suggests, causes the cells to get bigger and, if you're lucky, to produce recognizable changes in the nucleus, causing them to resemble owls' eyes. Poliovirus and other enteroviruses thrive in the gastrointestinal tract. In tissue culture they cause the infected cells to shrink. As they do, the cells pull apart from one another, leaving a stringy appearance in their wake. Some viruses, like respiratory syncytial virus, RSV, the most common cause of serious pulmonary infections in infants, produces syncytia in the affected cell sheets. These are multinucleated giant cells that display as their hallmark the presence of multiple nuclei that are left intact in the wake of the ravaging viruses. At the beginning of an epidemic of unknown etiology, no one knows which tissue cultures are best suited to provide a haven for the novel virus, so laboratories set up racks and racks of glass tubes, or trays and trays of small plastic flasks, which serve as the foundation for a wide variety of tissue culture types from a spectrum of animal origins. Traditional viral diagnostic labs keep in stock some mixture of cell cultures derived from human, simian, canine, rabbit, and guinea pig donors. The donor tissues are harvested from different organs, so the cells are genetically and morphologically dissimilar. In some cases, floating blood cells, such as lymphocytes, serve as the base for viral replication. The new virus can be grossly categorized based on the susceptibility of the cell cultures to infection and the cytopathic effect, CPE, produced, that is, the unique pathologic features seen through the microscope lens.

Once the virus gets going in cell culture, different outcomes may occur: huge amounts of progeny virus are shed into the tissue culture medium, leaving husks of destroyed cell skeletons behind; or the new viruses may fill the cells' interior and remain cell-bound, or cell-associated. The infected fluids or the dislodged, infected cells are centrifuged at ultrahigh speeds and the cell pellets, or viral heaps, are prepared for examination under the electron microscope. Viruses vary greatly in shape and size. Though they are very, very tiny, and are measured in nanometers, or billionths of a meter (that is, thousandths of a millimeter), their gorgeous symmetry and geometric patterns can be revealed to the human eye. Some viruses look long and wormy, while others have a perfect hexagonal geometry at their core, almost like a geodesic dome with sharp edges rather than curved. Some are bedecked in a garland or crown, like the SARS coronavirus. (See the photographs on pages 3, 47, and 143 for examples of viral architecture.)

For more than fifty years, researchers have analyzed the structure of viruses' genomes. There are only two choices—DNA or RNA. Today, in the era of highly advanced genetic analysis, one can sequence the entire chain of nucleotides in every RNA or DNA thread. The gene sequence precisely identifies each viral family or species, just as gene sequencing does for all forms of life on Earth, from bacteria to mammals. In recent years, gene sequencing has become so precise and so efficient that it allows us to pinpoint not only the species of origin of each sample, but also the identity of each individual within that species. The public knows this, thanks to the education it has received from dramatic police and crime scene investigations portrayed on television. Although we cannot determine the genetic identity of every single virus particle within a multitude, we can figure out if there are different clones or genetic lineages of viruses within a larger pool, where each clone is comprised of millions of individuals.

The multinational groups of scientists who demystified the ultimate source of the fatal respiratory disease that was spreading to three continents followed this sequence of identification steps. The SARS virus contained within bronchial lavage fluids grew in Vero E6 cells, a cell line of simian origin. Many of the Vero cells fused into multinucleated giant cells. One particular porcine virus showed similar growth patterns, but its genome sequence was clearly different from the sequence in SARS Co-V. Therefore, it could be safely said that pigs were not the animals responsible for starting the SARS epidemic. But if pigs had been shown to be the reservoir for SARS, no one would have been surprised. For example, swine viruses have often been the source of influenza pandemics. To further verify their results, scientists showed that antibodies that were isolated from the blood of SARS victims bound themselves to the infected Vero cells with remarkable avidity. This proved that viral antigens had already been incorporated in the Vero cell membranes. Equally important, the absence of anti-SARS antibodies in blood from persons in the general population proved that the SARS virus had emerged from its reservoir only recently and that only those who displayed symptoms or had been in contact with SARS patients had mounted an immunologic response. Antigen-antibody reactions such as these form the basis for a panoply of tests that are employed universally in diagnostic virology labs. Finally, electron microscopic examination of highly concentrated sputa or lung lavage fluids revealed a myriad of teeny crowned particles of exactly the right dimensions (see figure, page 3). Also, high concentrations of SARS RNA were found in a variety of clinical specimens, up to 100 million molecules per milliliter of sample.

Findings from the genetic sequence analysis proved that SARS-CoV was only distantly related to all other known animal and human coronaviruses. Because it was unique, it found easy prey within a virgin population, one that had had no experience with the virus and hence lacked any previous opportunity to mount an antibody response to temper the advances of this brand new renegade. Analysis of serial blood samples from its victims provided further proof that this novel coronavirus was the etiologic (or causal) agent of SARS. Antibodies were sometimes absent from serum samples taken during the very earliest stages of the illness, but later appeared in ever-increasing titers. The sudden appearance of antibodies to an infectious agent, a phenomenon called seroconversion, is considered definitive proof of acute infection, evidence that the host had never been infected previously but was most assuredly infected now. In another phenomenon, related to seroconversion and equally compelling, low antibody titers rise significantly, usually fourfold or more, during a brief period of observation. This increase demonstrates ongoing infection, accompanied by a specific and growing immunologic response.

Some patients had SARS-CoV in their stool. This was unsurprising, because many coronaviruses of animals were known to cause GI disease. However, researchers were surprised when SARS-CoV was recovered from the blood of the index patient in a small outbreak of SARS in Frankfurt, Germany. This finding proved that SARS-CoV had a viremic phase following its initial attachment phase, a period during which the virus moves from its portal of entry, in this case, the respiratory tract, through the bloodstream. This viral journey inevitably results in the dissemination of virus to multiple organs and portends more severe disease. Although SARS is quite rightly understood to be a severe lung disease, as its name indicates, its invasion of the gut and liver almost certainly contributes to the nasty clinical outcomes that befall its victims.

The SARS pandemic, the first in the twenty-first century, lasted only eight months. Despite worldwide fear that millions might be at risk, the total number of persons who fell ill was a little more than eight thousand. Twenty-nine countries reported victims. The case-fatality rate, the ratio between the number who died and the number who were infected was ten percent. While public health officials and epidemiologists wondered where SARS had gone, the more urgent question was when it would come back. Every year influenza returns and Ebola has returned to Africa on at least ten occasions. During its heyday, smallpox never went away; neither did measles, or rubella, or chickenpox, or polio, at least until effective vaccines were developed for all four in the mid to late twentieth century. SARS prompted great concern because of its clinical course, mortality rate, and high degree of transmission. In this era of globalization, the worry arises that international travelers could easily bring a disease like SARS from somewhere far away to your very home or hospital within a matter of days.

To analyze the concerns regarding a potential reappearance of SARS, in March, 2013, on the tenth anniversary of SARS' arrival in China, *Science* published an article called "War Stories." The authors asked: "a decade after the SARS outbreak, how much safer are we?" It was acknowledged that should SARS re-emerge, scientific advances in the fields of molecular virology and pharmacologic discovery would provide some assurance. The international cadre of virologists, through immediate application of both traditional and new techniques, could rapidly confirm the identity of the etiologic agent and could sequence its RNA genome. This would reveal whether the new circulating strain was identical to the former one and, if it differed, what these differences might predict in terms of the virus's potential pathogenicity and transmissibility. Also, the pharmacologists could begin to look through the troves of drugs on their shelves or currently in their pipeline. Could some old or newer agent be tested quickly *in vitro*, to assess its activity against SARS-CoV? Might the chemical structure of a drug used empirically ten years ago be tweaked, by adding a side chain here or there, to create a product that was a bit more effective? Of course, the new drug would have to be tested in animals and then in humans to make sure that it was not harmful or unacceptably discomfiting.

It was hoped that if SARS re-emerged in the Far East, local officials would be more transparent and quicker to report their findings than they had been a decade earlier. International public health agencies were still irked by the foot-dragging that hindered Chinese officials at various levels of the bureaucracy from announcing the outbreak in Guangdong province in time to stanch the initial spread of SARS to Hong Kong and beyond. During the epidemic a three-man international delegation of scientists who came to China to gather information had to wait nine days in Beijing before they were allowed to meet with the head of the Chinese CDC. They learned nothing and were then denied passage to Guangdong.

Even more recent evidence of possible bureaucratic delay during the MERS coronavirus outbreak in Saudi Arabia in 2012 has come to light. It took several months for news of the MERS outbreak to reach the outside world, and it took still longer before the reasons for this unfortunate time gap were fully explained. The first report of MERS in Saudi Arabia appeared on ProMED, an internet service that announces news of infectious disease outbreaks to the general public.

An Egyptian microbiologist reported the news while working at a hospital in Jeddah, Saudi Arabia. He had isolated the virus from a man who died of pneumonia. The lab worker sent samples of his newly discovered pathogen to a virologist in the Netherlands, who sequenced its RNA. He also notified the Saudi government and urged them to notify the WHO immediately. When the Saudi officials refused to take his advice, he announced his finding on the web. The Saudi deputy minister for public health denied that anything about MERS had been reported to him and was surprised to have learned about it when he saw notice of it on the internet. In the past, and perhaps still, officials in many countries fear that reports of a new disease would stigmatize their health systems and suggest that they were somehow unable to prevent the spread of a new disease or unclear about how to proceed, especially if that disease emerged from unhealthy or unseemly relationships between their citizens and animals.

These days most emerging diseases arise in the animal world; hence the name enzootic is applied to them. In most parts of the world the human and animal populations have close contact. The proximate contact becomes intimate when animals are used as a necessary source of protein, or of cash when the animals are hunted and sold illegally on the exotic animal market.

These illicit operations and dietary habits underscore the vital role played by information flow, trust between injured parties and investigative entities at the crux of international infection control. The problem arises not because of a lack of communication networks. They exist. At the outset of the SARS outbreak, the Global Outbreak Alert and Response Network was already up and active; it comprised laboratories and research institutes around the world. But it was so unsuccessful in engaging United Nations member states that they simply failed to report potentially critical events in a timely fashion. A number of governments felt their autonomy was being imperiled; officials in these countries decided they would notify the WHO when they were good and ready, when they were absolutely certain that they were dealing with a bona fide novel infectious disease outbreak that did not reflect any homegrown failure of hygienic or public health practices. The International Health Regulation, or IHR, a legal reporting framework, might have served as a gathering post for information, but its rules were last revised in 1969. The document promulgated practical steps for reporting outbreaks and handling diseases that could cross national borders, but it was outmoded in that it applied only to plague, cholera, and yellow fever. SARS was not one of these, so China and other affected countries were not obliged to report, and they didn't.

Recent reports from the field indicate that the situation has improved in recent years. In 2007 the IHR was revised. It now requires countries to report within twenty-four hours anything that "may constitute a public health emergency of international concern"—this stipulation included "unknown diseases." The WHO was also given the authority to investigate "informal reports" about disease outbreaks.

It's fair to say that deficiencies in infrastructure and financial support are the greatest impediments to a swift and perfectly orchestrated response to any infectious disease outbreak. Countries must put in place up-to-date and rapid surveillance networks, modern well-equipped diagnostic laboratories with trained professionals to staff them, and clinics with hospital beds that are equipped with the capacity to deliver intravenous fluids, blood products, and oxygen. The health facilities must also have access to an endless supply of infection control outfits (gloves, masks, protective eyewear, booties, gowns, etc.) and hand-washing facilities throughout. All this costs money—lots of money. Various forms of transport are required to carry patients, equipment, and supplies to the health centers, and their services only work efficiently when roads are cleared of rocks, ruts, and debris, and are straight and passable in all kinds of weather. The successful international response to the West African Ebola outbreak of 2014–15 came about only because some of these impediments were recognized and dealt with in a timely fashion. Amazingly, the outbreak ended without there being a vaccine or an antiviral medication that was proven to be effective. Excellent diagnostic tests were developed during the course of the epidemic, but they were not needed to change the spread of the disease, since everyone on the ground knew what Ebola looked like, including family members of those stricken.

Similar comments can be made about the response to SARS; there was no vaccine and no scientifically proven antiviral. Effectively, infection control and quarantine brought the disease to a halt in all the affected hotspots around the world. As noted in the March 2013 *Science* "War Stories" article:

> . . . in the end, 21st century lab science had little impact on the fight against SARS; the disease was stopped using 19th century hygiene measures. Diagnostic tests developed soon after the virus was isolated weren't needed to manage the outbreak, and the [RNA] sequence completed by Canadian scientists on 12 April, had little direct impact. Scientists launched programs to develop antivirals and vaccines against SARS—and some of the work is still going on—but they never came to fruition; the direct need for them disappeared in July 2003, when the epidemic was declared over . . .

If SARS should make a comeback, scientists everywhere will have to put on their white coats, gloves, and face masks and return to their biosafety hoods and get to work. They will already know everything they need to know about the virus's structure, its gene sequence and its clinical characteristics. They will already know in general what class of antivirals had been shown to dampen the growth of related coronaviruses in vitro. They will also know what elements of the virus anatomy are the likeliest targets at which to aim the immune system, especially its antibodies. This head start should make the response to the next outbreak quicker than in 2003. It would be wonderful if the first victims were reported to international health organizations as soon as they got sick.

CHAPTER 6

Miniature Masters of a Million Microenvironments

VIRUSES RULE!

> Now we have a very powerful animal virus circulating in Asia that is already moving into humans. It is very widespread. The virus has changed its characteristics over the past few months. It is now highly pathogenic to chickens. It is infecting an increasing number of species—we now know that cats are susceptible. And they don't just start coughing and sneezing, either. They die.
>
> Michael Specter, "Nature's Bioterrorist," *The New Yorker,* February 28, 2005

> That human health and the Earth's health are intertwined can sound like a truism—more of a bumper sticker or poetic truth than scientific fact. Yet a growing body of research suggests that disrupted ecologies may indeed produce more disease.
>
> Brandon Keim, *Anthropocene Magazine*, December 14, 2016

In a lucid and beautifully constructed article in a 2013 issue of the *Annual Review of Microbiology*, Brian Wasik and Paul Turner, professors in Yale's Department of Ecology and Evolutionary Biology, convincingly argued that viruses are "the most successful inhabitants of the biosphere." They describe and detail the various elements of viral superiority. Viruses have achieved colossal numerical abundance—they are estimated to outnumber all forms of cellular life combined. Viruses exist in diverse shapes and physical states, they display a startling reproductive capacity, and they adapt to innumerable macro-environments and small ecologic niches. They also "learn" to evade the host immune system. Most viruses ultimately develop resistance to the many antiviral drugs designed to lay them low. HIV is particularly deadly because it kills the very cells that were created to vanquish it.

As Darwin reasoned in the mid nineteenth century, all organisms on Earth change continually. Every part of an organism's structure is subject to change, as are its physiologic operations and biochemical reactions. Darwin didn't know about genes or DNA, the grand conductors of the symphony of inheritance and transformation, but he did understand that some changes benefit their hosts and some do not. The destructive changes curse their hosts with greater susceptibility to infection, disease, disability, or diminished adaptation to their home ground, while the beneficial changes bless their hosts with better connection to their little corners of the planet. An increased capacity to reproduce is perhaps the greatest gift offered by mutation, because it endows its bearers with the wherewithal to pass their good genes and characteristics to the next generation and the next and the next.

Viruses are merely tiny packages of DNA or RNA, encased in a protein capsule and layers of lipid with studs of sugar-coated proteins. They have been modified by genetic variation, gifting them with the sinister ability to grab onto some already-fixed molecule on the surface of their host cells which affords them entry to a fertile farm. The intracellular feast allows the virus to furiously replicate itself. The influenza viruses are masters of this technology and have gained notoriety because, even under tight surveillance, they keep changing and returning year after year in ever newer guises that may endow them with characteristics that result in pandemic bursts of disease and death. Oftentimes, newer strains of influenza replace one or more of the older ones, but sometimes they are merely added to the previous mix. Influenza is the only disease whose epidemic level outbreak heralds a rise in the mortality rate above that normally observed in nonepidemic years.

The damage potential of a virus is often referred to as virulence. Virulence is a mix of attributes that allows the virus to replicate more rapidly, to spread from host to host with greater efficiency, and to cause damage at the level of cells and, ultimately, whole organs which degrades the host's ability to survive. We worry most when a genetic change equips a virus with the ability to increase its host range, to spread from its indigenous species to another one which lives close by or which intrudes accidentally. This extension of host range is sometimes fraught with danger, either for the virus or its host. If the virus does too much damage to the host, it loses an opportunity to develop a long-term symbiotic relationship with it, particularly if the host dies very soon after infection. But if the virus forms a useful longer-term parasitic relationship, it can continue its mission of boundless replication and transmission. The host may be damaged just enough to cause long-term disability, but not so much that the virus loses an opportunity to commandeer the host on which it survives. This latter scenario is familiar to those of us who have witnessed the ravages of HIV. The virus produces billions of progeny every day or two, and the damage done to the human host is life-threatening. But the rhythms of growing disability can go on for years in the absence of pharmacologic intervention, while at the same time the infected man or woman continues to spread the infection by way of sexual encounters or even minor transfers of blood such as those that occur when sharing hypodermic needles. HIV also finds refuge in so-called privileged sites, such as the central nervous system, where the immune system cannot easily root out its prey.

The frequency of mutation-driven change varies greatly, depending on the composition of the viral genome—the RNA vs. DNA—and the differences in the coding sequences between one gene and another. In general, the effect of mutational frequency is inversely proportional to genomic size. A change that occurs in a small genome produces a greater effect because a greater percentage of the gene array is affected relative to its total size. It's been estimated that in RNA viruses, a small solitary change, or polymorphism, in the RNA chain occurs once in every thousand or hundred thousand replicative cycles. This may not sound like much, but viruses reproduce at great speeds, and hundreds of viruses, or more, inhabit every infected cell. Also, RNA-containing viruses such as influenza, or HIV, or the SARS coronavirus exhibit higher mutation rates because RNA viruses, unlike DNA viruses, are less able to correct errors that regularly occur in the course of viral replication. Not all mutations benefit the virus, but those that do often result in the qualities that bestow upon viruses their nasty reputations. Sometimes a mutation increases the capacity of a virus to evade the host immune response or its ability to change its host range, allowing the virus to move from subhuman to human primates, or from swine and fowl to humans. Influenza viruses are masterful jugglers on the stage of surface protein substitution. A couple of amino acid alterations that change the three dimensional conformation of either the hemagglutinin or neuraminidase molecules (represented by the H and the N designations that give influenza viruses their identity; see chapter 2) result in evasion of the neutralizing effect of pre-existing antibody. The resulting frustration of the host immune response frees the flu virus to spread like wildfire, even among populations that have seen more than their share of flu viruses in past years. Add to this ominous mix numerous genetic sorting reservoirs, like herds of swine and flocks of waterfowl, and you have the start of an epidemic that traditionally works its way east or west from Southeast Asia to the Western hemisphere.

According to Darwin's laws of adaptive evolution, the Earth has long been populated with interconnected and often closely related groups of animals. Entire ecosystems have evolved in similar fashion—there are deserts and hardwood forests, tundra and rainforests in both the Eastern and Western hemispheres. Having traversed the globe, humans have created a home in every part of the world. Among the most ancient self-generating inhabitants of our planet, viruses have also evolved in concert with their animal prey and shared environments. If a particular virus family is lucky enough to find a suitable dwelling place, it may also find a welcome host cell in a proper host animal where it will begin to construct its uncountable numbers of offspring, each an exact replica of its parent. Consider smallpox or measles, poliovirus, influenza, or HIV. They spread worldwide and infected humans on every continent, in hot and frozen climes alike. Some viruses have evolved armor so sturdy that they can survive for varying lengths of time on inanimate surfaces or in bodies of water, biding their time until an unwary animal or person touches the wrong thing, perhaps a table top or doorknob, or bathes in some contaminated reservoir or dirty stream. The polioviruses and respiratory syncytial virus, a major pathogen of infants and toddlers, represent species that are known for their ability to persevere in these ways.

William "Billy" Karesh, Executive Vice-President for Health and Policy of the Eco Health Alliance (formerly the

Wildlife Trust), holds to the theory that mammalian diversity, in concert with human population density, poses the greatest combined risk for the appearance of emerging infectious diseases, known as EID. Risky but perfectly natural human actions precipitate zoonotic spread. Hunting brings people deep into the homes of their prey; animal slaughter, butchering, and consumption of animal flesh bring hunter and prey into still more intimate alliances. International food markets, local markets, and international trade bring all of us into contact with people and animals from everywhere on Earth. There is a lucrative illegal intercontinental trade in wildlife and forbidden animal products. The United States and China have sometimes been accused of being among the worst violators in this tawdry business.

Climatic change, deforestation, and environmental degradation put pressure on human, animal, and viral populations, as they are forced to accommodate to the ecologic and genetic changes they encounter. Ultimately, mixed populations find a new way to survive in synergy, even harmony. Most often a specific event or new behavior generates the seed that blossoms into an emerging disease: a bat carrying a newly minted virus nips a goat that a farmer plans to slaughter the following week; a starving boy kills a monkey with his blowgun and brings it home for his family to feast upon. The monkey carries a virus in its blood that's already undergone numerous stages of mutation, allowing the virus to attach to epithelial cells in the boy's mouth and throat. A woman has an intimate encounter with a man who's incubating a novel infection that he acquired while ridding his home of a nest of infected rats and cleaning up their excrement. Billy Karesh estimates that a dozen or more animal groups share five or more viruses each with humans. These include birds, bats, pigs, primates, and rodents. So a little bit of transference here or there is to be expected. Insect vectors such as mosquitoes, fleas, or ticks are not discussed in this book, but imagine throwing billions of biting insects into the already swirling cauldron of animals and humans, and you see the problem of disease transmission can become unimaginably complex, but possible to solve nevertheless.

Viruses are inanimate, so one might not suspect that under special circumstances they are able to respond to changes in temperature or humidity. But there is experimental evidence, in particular data collected in Paul Turner's lab, to show that they do. Viruses mutate in order to adapt to the purely physical aspects of their environment. Their ability to mutate explains why viruses, unlike most living organisms, have found ways to populate every nook and cranny of the entire planet. Of the two nucleic acids, DNA is quite stable and somewhat resistant to decay. The outer coats or inner cores of some of the gastrointestinal pathogens, like noroviruses or rotaviruses, can withstand the desiccating effects of sitting on inanimate surfaces for hours, if not days. This gives them the ability to infect passengers on cruise ships or athletes in Olympic Villages, who intermingle in close quarters, all the while shedding these viruses as they travel to the sundecks, cocktail lounges, bathrooms, and training venues.

Dengue viruses, chikungunya virus, and Zika virus have gained notoriety of late because of their capacity to cause terrible disease outbreaks in the warmer and more humid places on Earth, but all three have shown a propensity to spread beyond the regions where they first arose. Some observers believe that these viruses may have begun to acquire the ability to adapt to cooler and drier environments, allowing them to spread into the continental United States from their usual haunts in South America and the Caribbean. Perhaps, but it's also possible that the known mosquito vectors which carry the viruses have also adapted to more temperate environments. Also, newer mosquito species seem to have acquired the characteristics that render them infectable. Esteban Domingo, of the Centro de Biología Molecular Severo Ochoa, in Barcelona, Spain, in a review article on mechanisms of viral emergence, favors a theory of "biological complexity." Viral emergence or re-emergence and the unanticipated spread of viruses to new hosts are paradigmatic of the "science of complexity." Domingo pays homage to the work of colleagues who define the science of complexity as "the study of those systems in which there is no simple and predictable relationship between levels, between the properties of parts and of wholes." Domingo goes on to explain:

> The emergence of viral disease involves several levels of complexity. The underlying level stems from the population structure of viral populations as they replicate in their standard hosts . . . The second level of complexity results from a network of environmental, ecologic, and sociologic influences that affect the probability that a potentially pathogenic virus comes into contact with a new host. A good number of such influences are subjected to indetermination.

I rarely come across words that I'm not familiar with, but in the weeks during which I wrote this chapter, I came across two—"abstemious" and "evolvability." To help myself remember the meaning of the first, I constructed the following sentence: "His abstemious habits drew him away from the dining table piled high with food and drink." Wasik and Turner provided their own definition for the second word in the scientific paper I referred to earlier: "We additionally consider whether viruses are advantaged in evolvability—the capacity to evolve, . . . in avoidance of extinction."

If Darwin had come up with this neologism, he might have sprinkled it throughout his masterwork as a device to emphasize the idea that forward evolutionary movement requires that a living organism, or in this case a virus, must retain the capacity to change and must use those changes to adapt to any given environment, that is, to evolve and hence, survive "in avoidance of extinction." We have long recognized that the environment is also subject to endless change. Geographic entities are prone to experience alterations in temperature or humidity, rainfall or tidal shifts, to which we now apply the broad term "climate change." Climate flux affects all flora and fauna within its particular boundaries. The viruses that parasitize the plants and animals must, in turn, adapt, that is, acquire a trait called "evolvability."

The word we use to describe a virus's inherent capacity to adapt is "fitness." In basic terms, fitness really means "replicative fitness," the capacity to produce enough infectious offspring so that some significant majority will survive in a particular ecologic niche, and, ideally, surpass in sheer numbers the members of other viral clones or families that compete for the same hosts and nutrients. Also, any change in structure or function must be permanent or nearly so, as it passes smoothly from one generation to the next.

Even higher animals that reproduce sexually have only a limited repertoire of genetic recombinational mechanics that either advance or retard the progress of individual lineages or species. The majority of these random genomic alterations or mutations result in harmful consequences; only a minority are beneficial. Viruses have three ways of juggling their genes: mutation, recombination, and reassortment. Small mutations come in three basic flavors—single nucleotide polymorphisms, or SNPs ("snips"), insertions, and deletions. In these latter two instances, one or a couple of nucleotides, the molecular beads that form the DNA or RNA strands, are added or subtracted as intact consecutive sequences. A second, grander style of mutation, recombination, is characterized by a veritable swap of a whole string of nucleotides on one gene with a similar length string on another gene. The influenza viruses are most adept at using the third mutational maneuver, reassortment, as its favored modus operandi. Because the entire influenza genome is divided into eight segments, each of which operates more or less as a single gene, the flu virus can swap one whole segment of one virus, or a whole gene, with a comparable segment in another virus; that is, it reassorts big pieces of genetic material. This happens only if two virus particles invade the same cell and spill all their genetic baggage into one pot before the swap occurs. The exchange of whole genes is a much bigger deal than the loss or gain of one or two nucleotides within a gene, because the resulting change in viral genotype and phenotype is huge. Such changes, or "shifts," are the progenitors of influenza pandemics.

Pandemics do not always provide unlimited benefit to the virus, even though in specific times and places the flu gains enormous strength by flooding the world with its progeny. In some cases, the flu is so deadly that many of its animal hosts will die while still burdened with its millions of intact viruses, having already prepared to go into the big, wide world. As a result, these unlucky viruses go to their graves along with their hosts, who will never be available again to serve as fertile ground for conquest. Furthermore, among the survivors of the flu, most will develop antibodies against the predominant strain. This immune response will provide some protection against infection, whenever that strain, or one closely related, reappears, a year or several years later, in its exact original guise. But influenza viruses are no dummies. Between pandemics, the remaining flu viruses continue to undergo minor mutations, called "drifts," which bestow upon them at least limited ability to infect new hosts, even those who may have had prior contact with their closely related viral ancestors.

Evolvability is not unique to viruses. All living organisms change and depend on this ability for their survival as a species. Viruses, although not officially living things, are deeply embedded in the biosphere and must get along with the hundreds of thousands of species on Earth that serve as the agents of their survival. After all, viruses cannot live on their own; they are entirely dependent on the hosts they parasitize. If the favored host is a mobile animal, it may migrate to an environment more favorable to its survival. In order to do that, it might have to alter its feeding habits or its body habitus, particularly if its new home is hotter or colder or wetter or drier than the earlier one. In the course of endless mutation, random genetic changes more favorable to the host's survival in a new corner of the biosphere may appear *prior* to its move, making it all the more likely that its adaptation will occur quickly. But it's just as likely that the beneficial genetic change occurs only after the move. In that case the adaptation must take place quickly, otherwise the host and its burden of endogenous virus will meet unhappy fates in unison.

The likelihood that a previously unrecognized viral pathogen will suddenly emerge in a susceptible human population is exceedingly small. There are so many variables that must come together in both space and time. This may be surprising, considering that in the past half century HIV, Ebola, Zika virus, and epidemic avian influenza have burst upon the world scene without prior warning. Perhaps the sudden appearance of these epidemics was anomalous, since all these viruses, or at least progenitors of each, have been around for centuries, if not millennia. After all, a defined series of transitions must occur in order for an animal virus to spark a pandemic. Sometimes these transitions occur quickly, sometimes slowly. A rudimentary description of the phases goes like this: First, the virus must be well-adapted to its reservoir animal host, allowing it to replicate in large enough numbers unimpeded by its host immune system. Second, the virus must find its way into those parts of the animal host which enable it to transit easily to its future human host. The human host must be able to eat, drink, touch, or inhale some bit of flesh or excreta (e.g., urine, feces, nasopharyngeal secretions) that carry lots of infectious viral particles. The virus must be able to find a portal of entry in its new human host. Third, that entry port must be graced with the right kinds of cells, perhaps mucosal epithelial cells or blood cells or nerve cells, that are already fitted with just the right kinds of receptors to which the virus can attach. These virological and biomolecular variables must be put in place at just the right time after eons of genetic modification and experimentation.

It takes only one person to start a pandemic. That infected individual must find himself in the company of other susceptible people whose body parts—noses, eyes, throats, skin, rectal or vaginal mucosa—are contiguous to the viral stream that emanates from the "index" person. The greater the concentration of virus in the stream, the more likely it is that a really "fit" virus will find a really willing host cell. This involves a bit of luck, of course. (Luck for the virus, not for the human.) The luckiest viruses are those that have acquired characteristics that foster their easy transmission from human to human. Respiratory viruses are particularly successful in this regard; they reproduce wildly in the nose and throat and can spread by coughing or sneezing or contaminating environmental surfaces through touch. (We don't often decontaminate our hands after we rub our eyes or scratch our noses or cover our coughs, even though we seem to be more and more inclined to decontaminate after a range of other activities.)

Viruses that we spread through sexual behaviors or that we shed in urine or feces are also among the most successful at precipitating outbreaks or epidemics. Sex is a behavior designed primarily to promote procreation and the future of the species, but it also serves to arouse pleasure and inevitably brings different mucous membranes into the most intimate of all encounters. In developing countries, the environment is often contaminated with excreta that carry disease pathogens, parasitic, viral, and bacterial. In this book I discuss only viral diseases.

Pandemic spread arises when person-to-person transmission becomes as efficient as spread of the virus among its original sylvatic or domestic animal species. In locales where population density is greatest this process is hastened. Crowding greatly enhances person-to-person spread. In some instances, epidemic viruses have found humans to be their final and only host species. Smallpox, measles, and mumps are examples of these "human only" viruses. Though smallpox has been eradicated, measles and mumps have not, despite the availability for about fifty years of preventive vaccines.

According to Jared Diamond, the American ecologist, biologist, and anthropologist, and his colleagues, most of the infectious diseases that occur in temperate climates are so-called "crowd epidemic diseases." They generally occur as brief outbreaks. If the disease is self-limited, not fatal, and therefore accompanied by prolonged immune responses, the reservoir of those persons susceptible to infection is depleted. The immune systems of the resistant population afford extended protection. In addition, in the absence of any animal reservoir, the epidemic dies out, at least temporarily. The disease cannot re-enter that same population until the number of new susceptibles rises to a level that allows entry by an outsider who carries the "bug" that serves as an agent of destruction. Measles is the perfect example of this kind of infection. In the pre-vaccine era, measles occurred in epidemic form once every two or three years, only at intervals when there were non-immune newborns and infants who entered the population in sufficient numbers to sustain a new outbreak. In recent years, measles outbreaks have occurred in the US only when an unvaccinated, foreign-born child with measles entered a local population in which there was a large enough number of unvaccinated youngsters to sustain an outbreak. These were groups, almost exclusively, in which religious conviction or unscientific beliefs led parents not to vaccinate their children in numbers large enough to give measles a ticket to ride. Sometimes, the hordes of children in places like amusement parks have contained large enough populations of susceptibles to spark an outbreak.

As Heraclitus noted in the fifth century B.C., "nothing endures but change." Viruses and all the life forms on Earth are changing all the time. Genomic change in the form of mutation is continuous. For viruses to behave in their usual harmful way, they have to coordinate their changes with changes all around them. Because of their sheer numbers and the small size of their genetic codes, viruses stand on the winning side. When trillions and trillions of genetic sequences in an infinity of viral genomes are changing all the time, every now and again a few lucky viruses win the game, take a great leap forward into the company of humans, and do more damage than can even be imagined.

While I was researching for this book, reading hundreds of articles from the medical literature, most recent but some centuries old, I felt increasingly compelled to try out a new word, to broaden the term "ecological niche," a moniker used frequently in discussions such as these. In an attempt to be more inclusive, I wished to gather into the term's meaning not only the flora, fauna, mineral, and meteorologic entities that comprise the myriad corners of our world, but also to include the sociologic, political, and economic variables that govern the deep relational patterns between viruses and animals that I've described. I've coined the term "omnilogical universe." Infectious diseases, both limited and epidemic, arise from an infinitely complex net of interactions between microbes and almost everything else!

Included in this "everything else" are "social contagion" and "disease-behavior interactions," terms introduced by

Alison Galvani, Ph.D., a professor at Yale's School of Public Health and an international authority on the power of mathematics to describe and, perhaps, predict patterns of spread of lethal infectious diseases. In Galvani's view the complex social networks among people, in which each individual is a node within the network, serve critical roles in the spread of contagious diseases and, conversely, in their prevention and control. During the most recent 2013–14 Ebola outbreak, for example, close relationships among family members and villagers affected the ways in which prohibitions against person-to-person contact were or were not observed. They also influenced the speed with which the afflicted found their way to nearby health facilities. If powerful people or celebrities promote or eschew specific medical interventions such as vaccination, uptake of beneficial treatments may be supported or defeated. When the president of South Africa condemned the use of antiretrovirals for persons with AIDS, and when religious figures in Nigeria vilified polio vaccine as a weapon of Western interference, the incidence of HIV and poliovirus infection soared. Communication within social networks affected the degree to which condoms were used to prevent HIV, and, on the other hand, the spread of the false belief that AIDS drugs cure the infection and thus give license to those infected to re-engage in risky sexual behaviors.

To amplify my "omnilogical" theory of disease evolution, I turned to Galvani's captivating disease-behavior model:

> . . . data on social networks are being collected to help formulate and test network-based disease-behavior models. Moreover, modelers are incorporating the insights of economists, sociologists and psychologists into disease-behavior models.

I'm taking the liberty of adding epidemiologists, biologists, and animal behaviorists to the mix of "insight formulators."

SARS spread from the Far East to Canada in the course of a day or two because air travel moves infected travelers so quickly. Ebola spread from the remote southeastern forests of Guinea to nearby Sierra Leone and Liberia due to a series of social upheavals, almost all traceable to extreme poverty and a tragic misallocation of the limited funds that governments failed to distribute. Too few regional health facilities and diagnostic technologies were available. Add to this a broken transportation system, a near absence of infection control supplies (gloves, masks, gowns, and clean water) initially, and poorly coordinated communication networks. In environments without sanitation, epidemic diseases spread geometrically. One infected person infects two others, they infect four or more, etc. In some well-documented instances, so-called "superspreaders" of Ebola virus passed infections to about eighty percent of the victims in the next generation of the epidemic. In one neighborhood of Freetown, Sierra Leone, one person was proven to have spread infection to twenty-four close contacts. As expected, they all fell ill in the course of one incubation period.

Bats and other wild animals have been known for centuries to carry pathogens to which humans are susceptible. A dearth of public health authorities and their local workers meant that they were unable to fully meet their obligations to the citizenry. They failed to advertise widely enough the hazards of hunting and eating bushmeat, even after the announcement that the very first fatality of the 2013 epidemic was a two-year-old boy in Guinea whose infection was caused by the Zaire strain of Ebola virus. The viral interloper arrived in Guinea in the saliva of a bat who set forth from another country in search of finer fruit. Even if warnings about eating ape or bat flesh reached the people, their abject poverty and longing for a favorite protein source may have caused the warning to fall on deaf ears. Apparently, hunger trumps the possibility of catching some unheard-of disease.

When poor people move to places nearby where agricultural opportunities are better and wild foodstuffs more plentiful, the move to improve one's lot also increases contact with all kinds of plants, animals, and microbes that have never been together before—new food for the humans creates more human prey for the "bugs."

Poverty sometimes drives people across unsupervised national borders into countries where jobs might be available. Shifting populations may carry shifting pathogens into human enclaves that lack immunity to the viruses they're about to confront. This scenario is merely a large-scale version of what happens when a solo traveler, often a child incubating measles, enters a community where the majority of its people are unimmunized and therefore susceptible to infection with a virus fraught with the possibility of serious complications. The wars and political strife that have left parts of West Africa decimated in recent decades also rendered useless much of the health infrastructure that even before the conflicts began was grossly underfunded and undeveloped. Conditions on the ground simply could not react expeditiously to an Ebola outbreak of epic proportions.

Many writers, and I include myself, struggle to find "le bon mot" and just the right turn of phrase. Sometimes we become prolix and salt our passages with too many big words and inept metaphors. We're not all poets or Hemingways. But sometimes the average Joe uses the simplest words to express striking ideas. I found a quote that states with perfect pitch precisely what I've tried to capture in the last few pages. It is from an article by Daniel Bausch and Marguerite Clougherty in the *Journal of Infectious Diseases*, 2015:

> The biggest problems at Donka are no electricity, no water, no equipment, no sanitation, and very high rates of infection.

The quote comes from Bintu Cissé, a midwife supervisor at Donka National Hospital in Conakry, Guinea, site of an outbreak of Ebola virus disease with nosocomial transmission.

On October 31, 2016, I was fortunate to attend a cautionary address and call to action by Jan Semenza, of the European Centre for Disease Prevention and Control, in Stockholm. The talk was entitled "Climate Change and other Drivers of Infectious Disease Events in Europe." We audience members knew that climate change posed a threat to the future of polar bears, seals, and coral reefs, but the idea that Europe is a "hot spot" for the emergence of communicable disease really spooked us. Dr. Semenza and his colleagues in the Section of Future Threats and Determinants have been tasked with the monumental job of anticipating future emerging diseases that arise as a result of business travel, tourism, migration, and refugee resettlement, all of which provide a platform for infections that are spread by insect vectors, rodents, water, food, and air. He did not mention camels, dogs, swine, and birds.

Under Semenza's guidance, epidemic intelligence officers focus on "disease monitoring," in which they identify unusual clinical cases, and "event monitoring," in which they analyze novel disease outbreaks. Semenza's early findings suggest climate change contributes to about sixty percent of all recent infectious disease events. Globalization is the second most frequent driver. Climate change influences foodborne and waterborne illnesses; poor water quality and breakdown of water systems also play a role. Epidemiologists want to discover which specific signals in the environment influence the incidence of animal-borne infection and what roles precipitation and sea surface temperatures play. What causes changes in the seasonal distribution of animal and plant species? As an example, it's been found that pathogenic vibrios, comma-shaped microbes like the cholera bacterium, proliferate best in warm, low-sodium waters, such as currently fill the Chesapeake Bay and Baltic Sea. Insect-borne viruses are not a focus of this book, but we do know that climate influences the activity of the mosquitos that spread West Nile virus, dengue, chikungunya, and yellow fever. Here is an example of the connection between insect-borne viruses and animals: In temperate zones, the median temperature in July influences bird flight patterns, and hence the incidence and distribution of West Nile disease, a disease spread by birds, particularly crows.

There is a relationship between hantavirus infection (the agent that set off the outbreak of respiratory disease in the Four Corners area of the American Southwest) and rodent population density. When rains are abundant, grass seed production soars, so deer mice, the purveyors of hantavirus, can gorge themselves gleefully. During warmer summers and autumns, tree seed production in Belgium increases and bank vole reproduction increases, along with the risk of hantavirus infection.

Plague is a bacterial disease, but its relation to climate is exemplary. The black rat harbors plague and fleas spread it. Except for limited outbreaks in parts of the American southwest and central Asia, there have been no plague epidemics since the early eighteenth century. But the times they are a-changin', along with the world's temperature. Black rats move indoors when it gets too hot outside; this increases their proximity to humans. (The infected fleas are carried inside free of charge.) According to Semenza, it's been projected that a 1° centigrade rise in spring temperature could herald a fifty percent increase in the prevalence of the plague organism in its reservoir hosts, which also include ground squirrels, prairie dogs, other wild rodents, and sometimes domestic cats. Although small clusters of plague cases are reported occasionally from New Mexico, Arizona, and southern California, we need not worry about a recurrence of its fourteenth century iteration. There are at least a half dozen antibiotics that kill the responsible bacterium, *Yersinia pestis*.

CHAPTER 7

It's Restraining Bats and Dogs

RABIES

Then finally there dawned on Pasteur a simple way out of his trouble: "It's not the dogs we must give our fourteen doses of vaccine," he pondered, "it's the human beings that have been bitten by the mad dogs . . ."

"How easy! . . . After a person has been bitten by a mad dog, it is always weeks before the disease develops in him . . . The virus has to crawl all the way from the bite to the brain . . . While that is going on we can shoot in our fourteen doses . . . and protect him!" . . .

And that night of July 6, 1885, they made the first injection of the weakened microbes of hydrophobia into a human being. Then, day after day, the boy Meister went without a hitch through his fourteen injections—which were only slight pricks of the hypodermic needle into his skin.

And the boy went home to Alsace and had never a sign of that dreadful disease . . .

The tortured bitten people of the world began to pour into the laboratory of the miracle-man of the Rue d'Ulm. Research for a moment came to an end in the messy small suite of rooms, while Pasteur and Roux and Chamberland sorted out polyglot crowds of mangled ones, babbling in a score of tongues: "Pasteur—save us!"

Paul de Kruif, *Microbe Hunters*, 1926

The word rabies means madness. It derives from the Latin *rabere*, to rave. The Greek word for rabies, *lyssa*, also means madness and is now the name applied to the genus Lyssavirus, to which the rabies virus belongs along with a few other close relatives. Descriptions of a disease which strongly resembles rabies date back several millennia to Babylonian times. They reappear in ancient Greek and Roman text, and, in the twelfth century, in the writings of Maimonides. Mammalian species other than Homo sapiens are known to be infectable. They include dogs, foxes, coyotes, skunks, badgers and martens, raccoons, mongooses, and civets; and lastly insectivorous and hematophagous bats.

The rabies virus belongs to the *rhabdoviridae* family. Like other members of its genus, rabies is a bullet-shaped virus. A helix comprising thirty to thirty-five perfectly spaced protein coils securely protects its RNA, and a spiked envelope forms the virus's outermost cloak. The rabies virus almost always causes a fatal form of encephalitis; it must get into the nervous system somehow and it must get "from the bite to the brain." The nerve cell membrane has one or more receptors for the virus. Once the virus attaches itself, the cell membrane encircles it, swallows it whole, and after a few more maneuvers, disgorges the RNA-rich central core, whose genetic sequences are transcribed into the five proteins which proceed to generate newborn baby rabies viruses. If the bite occurs on a limb, as it most often does, muscle cells become the first line of attack. Then the virus infects the nerves innervating the muscle spindles and begins to move centripetally along their far-reaching axons toward the central nervous system, beginning with the spinal cord. This fatal trip takes about two-and-a-half to three days or more.

No one knows how these miniscule virions find their way upstream via retrograde axoplasmic flow or how they negotiate the synaptic gaps as they move from one nerve cell to the next. The virus must wend its way along some four to six feet. It's thought that the virus replicates during its journey along the nerve cell so that on arrival it blasts the central nervous system with the largest inoculum possible. To further fortify the blast, rabies virus has found a way to enter both sensory and motor nerves, in a kind of neural double whammy. Bites on the face and neck are particularly deadly. The virus has to negotiate only a distance of inches before the fatal inoculum of fresh virus enters the complex web of nerves that lead directly to the mid brain or to that portion of the cord that traverses the neck.

There are differing theories concerning the action of the virus which finally results in its infamous neurologic dysfunction. Because there is little histologic evidence of neuronal necrosis, greater attention has been given to the hypothesis that the virus interferes with neurotransmission, particularly at the level of the synaptic gap. Ramped up inflammation certainly contributes to the damage. Some studies have demonstrated a thirty-fold increase in local nitric oxide production, and others have revealed a tampering with the endogenous opioid systems. The failure of the immune system to quash the onslaught results in a near hundred percent fatality rate. Because the central nervous system acts as a partial barrier to the salutary actions of the immune system it's considered a "privileged site," thereby permitting privileged pathogens to linger longer than they ought to, sometimes for a lifetime.

The pathologist partners with the diagnostic virologist in certifying the cause of death in victims of rabies. Hard to believe, but the gross appearance of the brain is usually normal. The microscopic examination reveals the true story. As in all forms of encephalitis, there are multiple signs of inflammation and vascular congestion, but the pathognomonic sign of rabies is the presence of Negri bodies, first described by Adelchi Negri in a German medical journal in 1903. Every med-

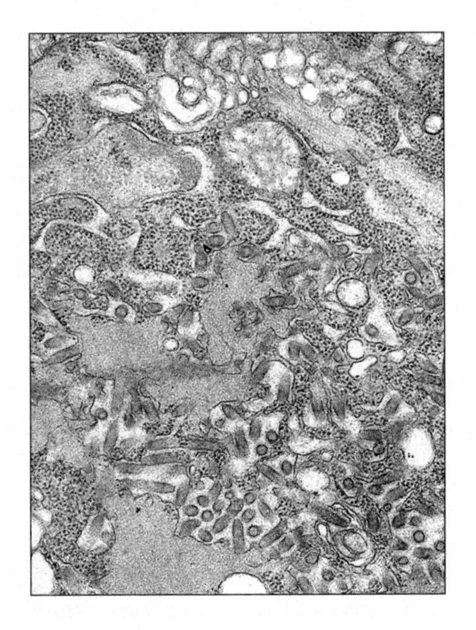

Electron micrograph of a cell infected with the rabies virus. The cytoplasm is filled with myriad rod- and bullet-shaped rabies viruses and pathognomonic Negri bodies, which are dense cellular inclusion bodies comprised of rabies virus structural proteins. Content provider: Dr. Fred Murphy, Centers for Disease Control and Prevention.

ical student learns about these cytoplasmic inclusion bodies, deeply stained round or oval objects averaging one to seven thousandths of a millimeter across, that are found in greatest concentration in the spinal cord, but also in the cerebellum and cerebral cortex. Super-ultrastructural analysis by electron microscopists reveals these Negri bodies to be comprised of the glommed-together cores of rabies viruses. Peripheral nerves are shorn of their outer coats of myelin, and when the rabies virus later reverses its course as it spreads centrifugally to tissues outside the central nervous system, it favors heart muscle, among other targets. The virus attacks the tiny nerve ganglions that innervate the cardiac atria before it slowly works its way to other parts of the heart.

Real diagnostic mavens sometimes offer the clinicians an opportunity to clinch the diagnosis in a most unusual way. They suggest doing a biopsy of tissue in the neck. A skin sample is removed by punch biopsy from the nape of the neck, just above the hairline. Thin sections of skin are stained with highly specific anti-rabies antibody tagged with fluorescein dye. Rabies virus tends to localize to hair follicles, and the antibody-bound virus particles emit an apple-green glow when viewed under the microscope.

It has long been known that rabies inexorably follows the bite of a rabid animal, most often that of a feral dog. But how, precisely, does the transmission work? It's unlikely that the teeth would serve as the proximate source of the virus, so it made sense to consider that the teeth act as a device for injecting the foaming drool that drips from the mouth of the furious animal. It was logical to incriminate saliva or some combination of saliva and other oropharyngeal secretions as the ultimate pool of infection, but this theory incites some alarm because oral secretions enter their surroundings in various forms and sizes—from visible droplets to showers of aerosols. They spread by way of coughs, sneezes, and drools. Was it possible that these infectious secretions might spread rabies to some unwary victim by something other than a bite? Could the gentle lick of a dog or raccoon, simply incubating rabies, transmit infection through a tiny rent in the skin? What about the invisible secretory emission delivered from a bat's mouth or nose, as it clings to a bedroom rafter above a silent sleeper? This last scenario has become something of a suburban nightmare, one that precipitates numerous calls to hospital emergency rooms and state health departments, but which, as far as we know, rarely results in bona fide disease. The possibility of this event occurring has been analyzed by bat behaviorists and animal virologists for years. Three sorts of "non-bite transmission" of variants of bat rabies viruses have been posited: aerosol transmission, transmission of true bites, little ones, that are unrecognized or minimized by the victim, and bites by an intermediate mammalian vector that was itself bitten by a bat. Even these scenarios are uncommon and thus rarely verified in the United States. Tens of thousands of people receive post-exposure prophylaxis for rabies yearly in the US for presumed possible transmission events that can never be verified.

In contrast, tens of thousands of people die from rabies in the developing countries of the world, the vast majority of whom succumb to feral dog bites. India appears to be a hot spot. In stark contrast, during the past decade or more there have been, on average, only two to three cases of confirmed human rabies cases in the US annually. Most were caused by variants of bat rabies virus, a half dozen by foreign dog variants and two by dog/coyote hybrid viruses. Despite the overwhelming fear of bats and their notorious association with rabies, it's estimated that less than one percent are infected. Of those bats actually submitted to state authorities for testing, about ten percent are positive for rabies, but they are predominantly sick bats, ones that have actually bitten their victims or have otherwise behaved strangely.

Bats hanging from bedroom ceilings, resting on sleeping victims' chests (really! it's been reported), have caused concern but not of the magnitude evoked by the trepidation surrounding bat-laden caves. Aerosol transmission of rabies is widely believed to pose a serious threat to humans visiting, exploring, or experimenting therein. However, apart from a single experiment designed to prove that the threat is real, but conducted under mostly unnatural conditions, there is little hard science, according to some experts, to lay the blame on aerosol spread of rabies. Nevertheless, proponents of aerosol transmission have repeatedly pointed to two cases in Texas, in 1956 and again in 1959, in which human rabies originated from exposure to bats in the Frio Cave. One victim was an entomologist who banded about ten thousand bats for migration studies months before his death. He also bled and inoculated bats with rabies viruses within weeks of his demise. The second decedent was a mining engineer who evaluated caves for possible mining of bat guano. Because both victims had a couple of skin lesions and scratches on various parts of their bodies, the "experts" could not definitively attribute their rabies infections to inhalation of virus-laden aerosols.

Many experts refuse to make definitive statements of fact until perfect controlled experiments conducted repeatedly under the most stringent conditions imposed by the scientific method produce results that appear reasonable to well-trained scientists. Consider the description of the conditions that existed in the Frio Cave in Texas in the 1960s. I found it hard not to believe that at least some episodes of aerosol transmission (or spread by multiple other avenues) sometimes occur in caves such as this:

> . . . at times perhaps 20 to 30 million bats occupy the cave . . . Human visits to the bat-inhabited areas of the cave are generally brief because of collisions with flying bats, attacks by arthropods, or inhalation of ammonia or other odors . . . Free-tail bats were attached as a continuous mat . . . up to 300 per square foot over areas of ceiling . . . The air below contained a light rain of urine drops, sticky guano pellets and mites . . . bat guano . . . varied in thickness from several inches to several feet . . . and bridged over holes in the floor . . . I have been contaminated with a near-continuous layer of mites . . . effect was a burning about the edges of the eyes, in the nostrils . . . ear cavities and the mouth edges were invaded also . . . (DG Constantine, Public Health Service publication, No. 1617, Atlanta, GA. Dept. of Health, Education and Welfare, 1967)

Inconspicuous, forgotten, or unnoticed bat bites are much more likely than aerosol contamination to be precipitants of the few human cases of rabies on record. Certain species weigh a fraction of an ounce, their teeth are tiny, and their bites are mostly attributed to spiders or stinging insects. People may not know that bats (or even skunks and raccoons) carry rabies. More worrisome are those persons who may present to an emergency room with early signs of rabies, having forgotten, over the course of a long incubation period, that they sustained a little bite weeks or months earlier. Or they remember, but fail to make the connection.

Unlike the developed countries of the world, the epidemiology of rabies is different in tropical and subtropical areas, including many parts of the Americas. The rich blood supply provided by wild and domesticated animals, particularly cattle, serves the dietary needs of the vampire bat (*Desmodus rotundus*). Each year, this bat is responsible for tens of thousands of cases of fatal bovine rabies and occasional cases of human rabies. Sadly, a related bat genus, *Lasionycteris*, serves as an emerging source of rabies virus for humans in the United States and Canada. There is concern that some rabid bats might recover from their initial disease and secrete virus in their saliva for months or years thereafter. For example, it's known that oropharyngeal fluids sometimes yield rabies virus even when the brains of such animals are sterile, in a kind of extra-neurological carrier state.

To test this hypothesis, a group of microbiologists, pathologists, and immunologists from Mexico and France conducted a series of elegant experiments to learn more about the patterns of salivary excretion of rabies virus by apparently healthy vampire bats. Two dozen vampire bats that lacked rabies antibody (that is, had never been previously infected) were captured in Mexico's *municipi el Naranjo*. They were isolated from one another and, in order to keep them well-nourished, they were fed blood from cattle that had never previously fallen prey to bat rabies. These bats were divided into two groups: one group was inoculated intramuscularly with a near fatal dose of vampire bat rabies virus. The control group was injected with sterile saline. All the animals were observed daily for ninety days for signs of neurologic disease—incoordination, poor feeding, tremor, flaccidity—and then weekly for 690 days. (On occasion, the signs of human rabies do not appear for months following exposure.) Samples of saliva were collected by oral rinse for two months. Less frequent samples were collected from healthy, sick, surviving, and control bats for 690 days.

Three-quarters of the infected bats died of rabies after a short period of diminished activity and anorexia. Some became tremulous and their wings became paralyzed. But none showed aggressive behavior. Three infected bats remained well even though all had developed antibodies to the virus, meaning they had been infected. Quite unexpectedly, no rabies virus was isolated from the saliva of animals that quickly succumbed to the full-blown disease! This result contrasts sharply with the disease's manifestations in other carnivores, such as dogs and skunks, whose saliva teems with virus when they're most aggressive. The scientists concluded that the bats with overwhelming infection died before much infectious virus found its way into saliva and before they could transmit infection to their usual victims. In contrast, they also found that rabies virus *can* be recovered from the saliva of infected survivors, but only intermittently and in the absence of neurologic impairment.

Some experts have raised the possibility that rabies virus may remain latent in brown fat or in the bone marrow, only to reactivate when the bats emerge from their hibernation period. Animal behaviorists also posit the theory that infected bats may transmit the virus to their mates by mutual grooming, food sharing, or contamination of the nest by a regurgitated blood meal, behaviors unrelated to biting. These conditions would increase the chance of reinfection in animals who had long since stopped shedding virus in the wake of past infection.

Most of the patients who die from bat rabies in the United States exhibit no history of a bite or scratch. This suggests that the most subtle exposure of a mucous membrane or abraded skin to virus may incite infection in a nearby nerve ending. In the United States proven cases of human rabies are rare, and the stories of their origin are mysterious. Between 1980 and 1993, only one case each was reported from New York, Texas, and California.

In August 1993, a fatal case of rabies infection in an eleven-year-old girl was reported to the New York State Department of Health. This was the first fatal case recorded in New York in thirty-nine years. The girl first complained of pain in the knuckles of her left hand. The pain increased and spread to her shoulder, after which she developed fever, severe muscle spasms of her left arm, difficulty walking, and hallucinations. She had a mild rash, but there were no signs of meningitis; nevertheless, she was given a course of both intravenous and oral antibiotics. Shortly thereafter, she refused to drink and had difficulty swallowing her own spit. Six days into her illness, her fever spiked to 105°F; she had some neck stiffness, but no other focal neurologic signs. The girl's spinal fluid showed few signs of inflammation; nevertheless, she was believed to have meningoencephalitis and was treated with another course of antibiotics and some steroids while being transported to a tertiary care center by helicopter. Upon arrival in the pediatric ICU she was noted to be alert and oriented to time and place. However, her pupils were unequal, an ominous sign of brainstem disease, and she developed respiratory distress, hypertension, tachycardia, and cardiac arrhythmias. She soon succumbed to irreversible cardiac arrest. Slices of brain obtained at autopsy were examined microscopically. There was profound swelling of all parts of the brain and some Negri bodies, the tiny telltale intracellular remnants of the innards of rabies virus. The little girl had no anti-rabies antibody in her blood, but bright apple-green rabies virus particles glowed in different parts of the brain stem, the midbrain, and the cerebellum, when stained with fluorescein-tagged rabies antibodies harvested from the blood of infected animals. To clinch the diagnosis, viral DNA extracted from the brain tissue was amplified by polymerase chain reaction, and the genetic sequence showed it to be a variant of the rabies virus most commonly found in insectivorous bats.

Intense interrogation of the family after the girl's death produced no useful clues to the origin of the fatal disease. Her relatives lived their entire lives in the deep woods of the Catskill mountains. There was no history of foreign travel, no history of close or distant contact with a bat, and no evidence of bat colonies on the property. There were lots of kid-friendly pets in the house—cats, dogs, gerbils, and rabbits—but none had been sick. The story was similarly unrevealing when all the nearby neighbors were questioned.

Rather unexpectedly, a little less than three months after the demise of the girl from New York, a Texan and a Californian died from rabies. An elderly farmer from east Texas was admitted to a hospital in Arkansas because of sudden onset of clumsiness, difficulty swallowing, forgetfulness, and confusion. The hospital emergency room records described hallucinations, uncooperativeness, increased muscle tone, and tremulousness. The spinal fluid was completely normal. It was suspected that the old man had had a stroke. He was artificially ventilated and pharmacologically paralyzed to calm his uncontrolled muscular paroxysms. Tetanus and rabies were then placed high on the list of more likely diagnoses, but no history of penetrating injuries, of animal bites, or of exposure to bats, raccoons, or skunks was elicited from the patient's family. The old man was encephalopathic and soon required mechanical ventilation, so he could neither hear nor speak. Five days after admission to the hospital he lost all his muscle tendon and brain stem reflexes. He failed to respond to painful stimuli. Life support was withdrawn. A sample of brain tissue obtained at autopsy was divided and sent to Arkansas and Texas. State-of-the-art molecular tests demonstrated that the virus isolated from the patient was closely related to strains of rabies found often in silver-haired bats in most parts of the US. However, the possibility that a rabid bat was *not* the culprit came to light when a neighbor of the dead man came forth unexpectedly, requesting rabies prophylaxis treatments. It turned out that the man had helped the decedent in caring for a sick cow with a strange undiagnosed illness three months earlier. Cows are known to be susceptible to rabies following bat bites, but there is no case on record of a cow having transmitted a bat rabies variant to its farmer friend.

In the same month as the unfortunate farmer, a man died of rabies in California. A native of Mexico, he was visiting relatives in California and presented to an urgent care facility with a several-day history of serious pain in his jaw, chest, and shoulder. It sounded like a heart attack. He also had a sore throat, anxiety, and vomiting. He was unable to drink or eat. He reported that he had had a spider bite on his left jaw; however, bite marks could not be found by the examining physicians. The patient aggressively refused all offers of oral fluids (hydrophobia) and his striking bouts of anxiety and dyspnea impelled his caregivers to give him injections of sedatives. When he calmed down, he was sent home only to return to a second hospital the following day in acute distress. He had a temperature of 103°F, high blood pressure, and hypersalivation. He spat uncontrollably as he staggered about the emergency room. Upon transfer to the intensive care unit, two nurses trained in the Philippines recognized almost immediately the classic signs of rabies. (Dog rabies is endemic in their home country.) With that critical insight in mind, the patient's family was asked more salient questions at which time they recalled the story they'd been told by their relative—he had been bitten on the left side of his neck by his neighbor's puppy. He told them that he cleaned the wound with soap and water, but received no rabies prophylaxis. Rabies shots were not routinely given to dogs in his part of Mexico.

Samples of the patient's saliva were sent to the Centers for Disease Control in Atlanta, Georgia. Molecular genetic tests revealed the virus in the sample to be related to strains of rabies often found in dogs in Mexico. It's become a kind of rule that rabies acquired abroad comes mostly from dogs and that rabies acquired domestically comes mostly from bats, skunks, raccoons, and foxes. One apocryphal indigenous case of human rabies followed a bobcat bite in California in 1969.

As clinicians, we are told stories by our patients every day. As the previous stories illustrate, each is unique and some are downright remarkable, even memorable, and many of us feel an urge to relate these amazing stories to one another. We are reminded on a daily basis that our diagnostic exercises must take into account not only the details of human behavior—what we eat, the people and animals with whom we cohabitate, the places we travel to, our modes of transportation, the drugs we take, our leisure activities, etc., etc.—but also where we live and where we work. Doctors are a nosy lot, and we depend on our patients to trust us and to forgive us for our probing inquiries. We teach our students that it is the patient's story that serves as the seed that ultimately germinates in a trusting, empathetic relationship.

There are many important facts the doctors, nurses, public health workers, and epidemiologists learned from the three patients who died of rabies in New York, Texas, and California in the summer and fall of 1993. They learned a lot about rabies not only from skillful examination of brain tissue post-mortem, but from speaking at length with relatives and neighbors. The patients themselves were too ill to offer up any salient facts. Because of communication barriers, more than half the cases of rabies diagnosed in the United States in the thirteen years prior to 1993 were only diagnosed post-mortem.

Human cases of rabies are uncommon in the United States and other developed countries because canine rabies prophylaxis through near universal immunization is part of routine public health practice. Also, concentrated rabies virus antibody preparations and post-exposure vaccines are widely available. Many cases diagnosed in the United States are actually acquired in foreign countries. Most infectious disease specialists have never seen a case of rabies in the course of their careers, but it comes up in discussions we have at the bedside or in our conference rooms when we consider the differential diagnosis and empiric treatment of patients with rapidly advancing encephalopathy and neuromuscular dysfunction. Only the occasional patient with rabies who presents with the classic symptoms of hydrophobia, copious salivation, and frenzied outbursts would serve to spark our memory of rabies stories we read about as youngsters (viz. *Old Yeller*) or saw in movies (viz. *Rage*, starring Glenn Ford).

The public health officers who investigated the three cases in 1993 reminded us that even rapid institution of treatment for rabies patients fails to alter the frightful, fatal outcome. They also taught us that each case must trigger the identification, interrogation, and possible treatment of all persons who had been in close contact with the index patient and who might have been exposed to his infectious saliva or who may have been exposed to the same rabid animal as the person already afflicted. Such persons need to be assessed as potential candidates for post-exposure prophylaxis. In the wake of the three American cases, epidemiologists also reminded us that human and animal rabies has been reported from all forty-eight contiguous states and that the silver-haired bat is most often accused of being the responsible animal. Scientists at the Center for Disease Control and elsewhere who have studied this particular bat have described a solitary, migratory species that prefers old growth forests. But in extensive surveys of multiple bat species designed to quantify the prevalence of the rabies virus, the silver-haired bat, despite its bad reputation, is among the least likely to carry the virus in the wild. Also, the particular viral variant most often found in the silver-haired bat is rarely found in other bat species or in those terrestrial mammals known to carry rabies. These facts leave clinicians and public health officers with the unenviable quandary of having to consider all bat species as potential carriers of rabies. As a consequence, infectious disease specialists are wont to recommend post-exposure prophylaxis for all persons bitten or scratched by a bat, and sometimes for those who have spent the night in a closed room that was shared with a bat, any bat, which nested upside-down on a bedroom rafter. We have been made to believe that bat saliva is infectious, and we are warned that bat bites may be more difficult to recognize than those of terrestrial animals, raccoons or skunks, for example. We are also asked to make every effort to exclude bats from our homes and to keep them away from our cats and dogs.

We all knew that there was one good thing about rabies—it could *not* be transmitted between humans! Or so we thought until we read a brief report in the *New England Journal of Medicine* in March 1979. The article was titled, "Human to Human Transmission of Rabies Virus by Corneal Transplant." In their report, the authors were wise to remind the reader that transmission of Creutzfeldt-Jakob disease, a terrible degenerative viral disease of the central nervous system, had been shown previously to have resulted from a corneal transplant. So there was already a precedent for the idea that spread of a viral disease by way of corneal transplant was possible. The transmission of rabies virus by this route was something brand new.

A previously healthy thirty-nine-year-old forester-rancher suffered the sudden onset of lower back pain and numbness over the lower part of his chest. The next day he became unsteady on his legs and afterward experienced weakness in all four limbs. When he was admitted to the hospital six days after disease onset he was able to walk only with assistance. The next day he could not get out of bed and the day after that he had trouble breathing and swallowing. He was intubated when he could no longer handle his oral secretions. Over the next four to five days he developed weakness of his eye and facial muscles. Progressive weakness of all four extremities was followed by loss of all of his deep tendon reflexes, such as his knee and elbow jerk. A test for myasthenia gravis was negative. On the sixth hospital day he suffered a cardiac arrest, but was successfully resuscitated. However, over the next two days he became comatose and his pupils became unresponsive to light. His electroencephalogram was markedly abnormal. He became hypothermic and hypotensive and died on the sixteenth hospital day. No diagnosis was forthcoming and the cornea from his right eye was harvested and transplanted to a thirty-seven-year-old woman.

Four and a half weeks after her transplant the woman complained of pain behind the eye, in the back of the neck, and across the shoulder. The symptoms progressed at a steady pace; the headache involved half her head, and there was facial numbness that soon spread to the right arm. She soon had difficulty forming her words and writing sentences properly. Then came swallowing difficulty. She became dehydrated. Despite all these troubles, the woman's intellect and cognition tested normal and, surprisingly, a CAT scan of the brain and spinal cord were also normal.

Beginning eight hours after admission to the hospital, the woman became more and more inarticulate. She was placed on ventilatory support when her breathing became labored and she could no longer swallow. She became quadriplegic and comatose and suffered cardiovascular failure. She died on the fourteenth hospital day, fifty days after she received her corneal transplant.

The mysterious relationship between these two remarkable clinical scenarios evoked many pathologic and virologic studies. While they were alive, the two patients remained puzzles, but the luxury of time and the contribution from additional experts were expected to yield a diagnosis that made sense. Standard light microscopy revealed severe meningoencephalitis in every part of the brain and spinal cord. Inflammatory white blood cells gathered in the recipient's optic nerve. Pink-staining inclusion bodies, clusters of viral particles or pieces thereof, were found in neurons throughout the nervous system. Rabies virus particles were seen in the temporal lobe of the cornea recipient and suckling mice developed rabies when they were inoculated with tissues taken from both the donor's and recipient's eyes. It seemed likely that rabies virus in the transplanted cornea traveled to other parts of the recipient's nervous system by way of the optic nerve or the trigeminal nerve, a large sensory nerve that divides into three branches as it supplies all parts of the face and part of the eye.

This very odd story reveals that rabies is rarely considered in the diagnosis of central nervous system afflictions that begin, as they usually do, without the classical signs of rabies and without a history of animal bite. Early on, some cases of rabies present with signs consistent with Guillain-Barré syndrome, a much more common and easily-recognized condition. And meningoencephalitis has many more obvious viral etiologies, including herpes simplex, West Nile virus, and other mosquito-borne infections such as Zika, LaCrosse, and the equine encephalitis. There are potential risks associated with the harvest for transplant of corneas from any donor with presumed viral disease of the central nervous system. The brain and the eye are interconnected. Embryologists tell us that much of the eye is an outgrowth of the forebrain.

There's a word association test that psychologists like to use in their practices. They claim the answers give them insights into how their patients' minds work. If I were to announce the word "rabies" to an average man in the street, I predict that most would respond with the word "dogs," or maybe "bats." However, had I been asked to associate a word with rabies in my late childhood or adolescence, I might very well have responded with the words "Louis Pasteur," not because I was some sort of savant, but because I was often exposed to Pasteur's story and those of Darwin, Newton, Einstein, Galileo, and others by way of the picture books and the more mature chapter books I read at home and in the library. While my older brother was attracted to illustrated books about astronomy and evolutionary biology, I liked stories about men of science. I remember reading *Microbe Hunters* by Paul de Kruif over and over again. The tales about Leeuwenhoek, Metchnikoff, and Walter Reed were great, but two separate chapters were devoted to Pasteur. I wondered why de Kruif paid special attention to the French scientist, until I went back and re-read the chapters. I think I understood why de Kruif and multitudes of others venerated this unique man of science. His interests were legion, his accomplishments were dazzling, and his role as a mentor put him in a category all his own. No exposition about rabies would be complete without reviewing the contributions made by Louis Pasteur. His creation of the first rabies vaccine came late in his life.

Pasteur was a chemist by trade. In the late 1840s and 50s the science of crystallography consumed his interest. His debunking of the theory of spontaneous generation and his work on fermentation are credited with saving the French wine industry. He studied silkworm blight and proposed ways to control it. Wine, vinegar, and beer production all depend on the behavior of microbes, so in studying these liquids, Pasteur inadvertently became a microbiologist and, soon thereafter, a student and later a proponent of the "germ theory" of disease, the practical consequences of which became the foundation of all Pasteur's research during the last twenty years of his life.

Before he got to rabies, Pasteur devoted himself to the study and prevention of anthrax, chicken cholera, and swine erysipelas. Pasteur was an exemplar of "bench-to-bedside" research. His pragmatic concerns drew directly from everything he learned in the lab. Everyone who drinks milk knows the word "pasteurized." You have to kill the microbes that make milk go sour and that cause bovine tuberculosis (in humans) and brucellosis. Rabies was Pasteur's ultimate pragmatic concern, but an odd one to be sure. Rabies is not and never was a great killer of humans. But its connection in legend with mad dogs racing through streets leaving in their wake drops of foaming drool gave it a place alongside vampires, werewolves, and gnarly witches. Pasteur himself sometimes recounted a disturbing incident from his youth in which a mad wolf ran through the streets of Arbois, the home of his youth, attacking shepherd boys, whose terrified cries mingled with the blood-curdling yelps of the rabid animal. Pasteur the romantic, the artist, the savior, believed that he could vanquish the terror and suffering that rabies delivered.

True to his fastidious and compulsive way of doing things, the now sixty-one-year-old scientist began a calculated series of experiments that culminated in the creation of the world's first rabies vaccine. But even this great scientist made some mistakes along the way. In one study he attributed the disease to the microbes he saw in rabid dog saliva. They turned out to be bacteria, and when he finally thought to sample the mouths of perfectly healthy dogs, he found the same bacteria, microbes that are simply part of the normal flora of all canines.

He made an even greater error when he decided to travel the same path he had trodden in creating his anthrax vaccine. He chose to invent a prophylactic vaccine, one that would protect all dogs in France from the dread disease by vaccinating them with small doses of rabies virus, enough to induce a protective antibody response, but far less than would be needed to cause clinical infection. Pasteur was a genius, but he had forgotten that there were at least a hundred thousand dogs in Paris, and certainly more than a million in all of France. Obviously, Pasteur's plan was impossible to enact. He did not have a factory operation, nor did he command an army of vaccinators.

Plan B was better. After many experiments, each involving dozens of dogs and rabbits, he found that some healthy dogs that had been bitten by their rabid cagemates could be protected from the onset of disease by giving them doses of a new rabies concoction that he had created out of mashed bits of air-dried rabid rabbit spinal cord tissue and that served as the essence of Pasteur's vaccine. He later learned that injecting the exposed but still healthy dogs with fourteen daily doses of vaccine, each aliquot a bit more virulent than the previous one, produced nearly perfect results. Only one or two of the dogs that had been bitten by rabid cagemates showed signs of rabies.

Pasteur was ready to experiment on himself. He would expose himself to the bite of a rabid dog to see if his vaccine worked. "I am beginning to feel very sure of my results," he wrote to an old friend. Serendipitously, Little Joey Meister appeared at the door of his laboratory in July, 1885, and relieved Pasteur of his foolhardy plan to experiment on himself. The experiment was, instead, performed on a nine-year old Alsatian boy whose arms were covered with festering red wounds, acquired while fighting off a mad dog. Joey got his fourteen daily injections of Pasteur's infectious suspension of rabbit spinal cord tissue. The pricks of the needle were no big deal; Joey felt great, returned to Alsace and never got sick. Neither did Jean-Baptiste Jupille, who was vaccinated after having been bitten by a rabid dog that charged him and six younger shepherds in a meadow while they were minding their sheep. The young lads were lucky, but Pasteur was even luckier. He didn't know what was in his very crude "vaccine." He didn't know anything about viruses. He didn't know how many billions were floating around in his miraculous soup, and he didn't know if they were alive or dead or some unstable mixture of the living and the dead. These were the good old, bad old days when you could get away with human experiments such as these. If Master Meister had died of rabies, it's possible that Pasteur's then glorious career would have come to an end. His election to the Académie Francaise in 1882 would have been the last of his great honors. The excellent annuities that came to him annually by way of his patron, Louis Napoleon, might well have dried up.

Despite the limits of his knowledge, Pasteur's experiments succeeded because rabies has a very long incubation period. "Post-exposure prophylaxis" is only practical if the interval between exposure to the virus and vaccination is long enough to allow the victim time to get from the site of the bite incident to the person with the syringe of vaccine. In Pasteur's day, people traveled by horse-drawn carriage, steam train or ships, trips that took days to weeks. It usually takes longer than that for the virus to get from the bite to the brain. Other viruses make their deadly trek to vital organs more quickly. Nowadays we most often use "post-exposure prophylaxis" to prevent HIV infection among health workers who suffer accidental needle sticks or for persons who had sexual relations with a partner who is or may be infected with HIV. In the case of HIV, we recommend that the person at risk receive a hefty dose of antiretroviral drugs as soon as possible, preferably within twenty-four hours. Unlike the rabies virus, HIV gets into the bloodstream quickly and begins to infect cells in the skin, the mouth or the genitals in a matter of hours.

Louis Pasteur became an instant hero in Europe before his fame spread to America. There were a few brief notices of his work, easy to miss, in the *New York Times* and the *New York Herald*. As always, bad news travels faster than good, and Pasteur's name ultimately came to everyone's lips thanks to a tragic event that took place in Newark, New Jersey. A wild dog ran through the streets and bit seven other dogs and six children. A local physician who had heard about Pasteur wrote to the newspaper urging the public to collect enough money to send the unlucky boys to Paris for a "new treatment." Pasteur was wired and agreed to see the stricken children: "Si croyez danger, envoyez enfants immediatement." The printing presses in the cities of the American Northeast spilled out stories about hydrophobia and how to apply a variety of folk remedies. There were cries to rid the streets of "stray curs" by shooting, poisoning, or drowning.

The boys' voyage to France aboard a freighter, outfitted with a makeshift "hospital room" and filled with donated warm weather gear for a frigid north Atlantic trip, gripped the public imagination. But these lovely human interest stories were soon mixed with more serious fare: reports of other noteworthy scientific discoveries, interesting but still nascent medical advances and various theories about vaccination, its practical aspects and its putative benefits.

The miracles performed on two boys in France and on six boys from New Jersey awakened the public to the practical consequences of scientific inquiry, to outcomes that were literally life-saving. People began to believe that similar miracles might be applied to the much more common plagues of mankind: tuberculosis, diphtheria, and whooping cough, for example.

The rabies vaccine was a "breakthrough" because, according to the historian Bert Hansen, it was characterized by five necessary attributes: "the advances must be seen as large, sudden, useful, already realized rather than just potential and of interest to a wide public . . . it is a social phenomenon; it is something widely noticed at the time." The polio vaccine, the eradication of smallpox, and the identification of the AIDS virus were, in our time, similar breakthroughs. The therapeutic nostrums of Galen and Hippocrates could be forever cast aside. The people wanted proven cures and preventatives. In the twentieth century and in the twenty-first, the public has been rewarded in spades, thanks to the largesse of government and industry, who wisely recognize and take advantage of the scientific breakthroughs that with some regularity come from university laboratories.

CHAPTER 8

Behold the Earth . . . One Mighty Blood Spot

EBOLA VIRUS DISEASE

> In June 2002, a gorilla carcass was found 15 km west of the sanctuary . . . In the next 4 months, we found 32 carcasses. Twelve of the carcasses were assayed for ZEBOV, and 9 tested positive. From October 2002 to January 2003, 91% (130/143) of the individually known gorillas in our study groups had disappeared.
>
> *Science Magazine*, December 8, 2006

Laser-like attention was brought to bear on viral-induced hemorrhagic fever in 1967, when simultaneous outbreaks of infections caused by novel viruses erupted in Marburg, Frankfurt, and Belgrade. Lab workers in all three cities fell ill. They had been dissecting kidneys from African green monkeys in order to cultivate simian tissue cultures in which to generate new virus stocks. Some of the healthcare workers who attended to the needs of the sick laboratory technicians also became sick. In all, thirty-one persons sickened and seven died after a horrific week of fever, internal bleeding, peeling skin, and shock. A virus was ultimately isolated from blood samples

---

The chapter title comes from a line in Nathaniel Hawthorne's story "Young Goodman Brown."

that were inoculated into guinea pigs and tissue cultures. The infectious agent was named Marburg virus after the city in Germany where its features were first described. The virus was novel in both its shape and protein composition. Many of the Ugandan monkeys who donated their kidneys for these experiments died of a hemorrhagic disease similar to the one afflicting their prosectors. So did all the African green monkeys that had received kidney cell transplants from their fellow simians.

After this scary preview, an uneventful period of eight years ensued that was finally punctuated by a cluster of three additional Marburg virus cases; two victims were South Africans who had recently returned from Zimbabwe and the third was a nurse who cared for one of the travelers. In 1980, a Ugandan died from Marburg disease and his attending physician was infected. A handful of unrelated Marburg patients appeared in East Africa during the 1980s. One was a gentleman who had investigated a bat cave right before he fell ill, raising the intriguing possibility that bats might serve as a reservoir for Marburg virus.

Marburg disease turned out to be a mere precursor to its far more devastating hemorrhagic cousin, Ebola virus disease. In the late 1970s, the simultaneous circulation of these two similar agents raised the questions, which culprit was doing what? Which patient was infected with which virus? Cross-reactive antibodies to both viruses in blood samples taken from many inhabitants of the affected areas caused consternation among epidemiologists in the field. Was there reason to pay greater attention to one virus rather than to the other? Yes, apparently there was, as was soon discovered.

In 1976, large outbreaks of hemorrhagic fever occurred simultaneously in Zaire and Sudan. More than 550 cases were reported, with an eighty percent mortality rate. Ebola virus, named after a small river in northwest Zaire, was isolated from patients in both countries. Electron microscopic inspection revealed that EBoV was morphologically identical to Marburg, but newer laboratory assays showed that antibodies to the two pathogens were, in fact, distinct and that tests could be used in the laboratory to discriminate between the two viruses, even though each produced the same clinical manifestations.

All simian and rodent species suffer the same devastating effects of Ebola virus disease after an average incubation period of four to ten days. During this period, EBoV replicates wildly in those organs of the body that contain large numbers of lymphoid cells, including the liver, spleen, lungs, and lymph nodes. When clinical symptoms appear, portions of these target organs have already begun to die. There is widespread necrosis and hemorrhage in the liver and in the myraid islands of lymphatic tissue in the gastrointestinal tract. Unstoppable GI bleeding ensues. Patients vomit up red blood and their stools turn black. As liver cells die one by one, the enzymes they carry pour into the bloodstream. Reduced manufacture of clotting factors in the liver along with malfunction of platelets and cells that form the inner lining of blood vessels ultimately leads to edema, hemorrhage, and shock. A maldistribution of the body's fluids and capillary leakage floods the lungs and other organs. The kidneys lose their capacity to ably filter the blood's waste. The cells in affected organs become veritable factories of viral genesis. The cytoplasm fills with newly manufactured clusters of viral core proteins, and the host cell membrane is studded with fully-formed, wormy, infectious Ebola babies, a virtual Medusa's head, ready to shed its noxious progeny.

Wormy or perhaps snaky are excellent characterizations of the Ebola virus. Textbooks sometimes describe its shape as "unique." Although its form once reminded virologists of a rhabdovirus, of which bullet-shaped rabies is an example, the taxonomists created a separate family to house the Marburg and Ebola viruses. They have a taxon or genus of their very own, the filoviruses. There are only six members of this select group: Marburg virus on the one hand and the Sudan, Tai forest, Reston, Zaire, and Bundibugyo subtypes of Ebola virus on the other. The Marburg and Ebola viruses do not share any proteins and, unlike most other RNA viruses, genetic variants do not emerge in nature as rapidly as in other members of the family.

The filoviruses are filamentous; their threads are very long (for viruses, that is). Sometimes they branch, and occasionally they bud or twist into pretzel shapes or rounded numeral-like conformations. Even though they're long, only a small segment, less than one-tenth, is needed to initiate infection. The function of the other ninety percent is being studied. Its outer shell is stolen from the host cell's membrane as the viral core enrobes itself when it exits from the cell's interior. Dozens of these viral cores clump together in the cell cytoplasm, forming "inclusion bodies" so large that they can be observed through a regular light microscope when properly stained. The details of EBoV's entry into the host cell is largely unknown. Entry sometimes occurs in the presence of the antibodies that are meant to neutralize it. This is not good for its victims.

The acute signs and symptoms of Ebola virus disease are hideous, but there are chronic manifestations of the disease that, though less repugnant, are equally important because of the resulting long-term disabilities. Many of these sequelae were first recognized in the wake of the outbreaks that occurred fifteen to twenty years ago. The most common, affecting at least half of all survivors, were chronic headache and backache, arthralgia, anemia, and various skin rashes. Hearing loss, neuropathy, alopecia, and weight loss also occurred, but less frequently. The complications resulting from the ocular diseases were of greatest concern because they may lead to blindness.

Emergency response teams organized by Médecins Sans Frontières proved that the most severe long-term complications of Ebola could be interrupted by personnel working in compounds that offered care to survivors. Some, such as those in Freetown, Sierra Leone, have performed up to a thousand consultations. Skin disorders, hair loss, arthritis, and ocular complications were among the afflictions most often encountered; some appeared soon after discharge from the acute care hospital. Patients who came in with red eyes often went on to develop uveitis, an inflammation of the pigmented layer that lies below the cornea; they needed to be admitted to the hospital emergently, where they stayed for over a week while receiving anti-inflammatory drugs and medicines to manage elevated ocular pressure. Despite great effort, some patients went on to develop cataracts.

At each visit, survivors of Ebola received psychological support and family planning advice. They were also taught how to prevent sexual transmission of the virus. Others were tested for HIV, which is prevalent in the area. Teams of healthcare providers went to local communities to educate villagers about Ebola and its mode of spread and to provide support, encouraging community reintegration and destigmatization of Ebola survivors. (Not infrequently, family, friends, and neighbors were unwilling to shake hands with or hug a survivor.)

In some outbreaks of Ebola, the case fatality rates exceeded eighty percent. At the other end of the spectrum, a minority of infections remained subclinical. In each cluster, it was observed that secondary and tertiary cases were often milder than index cases, suggesting that the severity of infection might lessen as it passes from one human victim to the next. Careful observation of the earliest cases proved that Ebola could spread by close contact, including sexual contact, and that blood, even in tiny amounts, was a source of transmission, as was later verified when Ebola virus returned with a vengeance in 2013. Hospital workers and their families were soon shown to be at greatest risk.

In the early years following the discovery of EBoV, its source in the wild was eagerly sought, but none was found. No trace of the virus or antibody to it was detected in samples from thousands of feral animals, including hundreds of monkeys. Reactive antibody, detected in nearly twenty percent of people living in the affected areas and tested in the earliest days of the epidemic, turned out to be non-specific. The antibodies reacted with more than one virus. Even guinea pigs living in affected areas had blood that produced false-positive antibody test results. These preliminary but unsatisfactory results only strengthened the urgent need to find the true source of the virus in nature.

Ultimately, random observations by locals and intense investigations by epidemiologists combined harmoniously to uncover the feral source of Ebola outbreaks that occurred in the mid to late 1990s and again in 2013–14. Beginning around 1995, Gabon and Congo recorded self-limited human clusters of Ebola virus disease. In each instance, human disease was accompanied by the accumulation of gorilla and chimpanzee carcasses in the nearby forests. These ape die-offs were a cause for alarm among animal conservationists, and their concern only intensified when in 2002 the Zaire strain of Ebola virus was identified as the cause. Scientists counted almost five thousand gorilla deaths, over ninety percent of which were due to Ebola. To the extreme dismay of workers at the Lossi Sanctuary in the Republic of Congo, Ebola spread southward during late 2003, leaving in its wake the death of nearly ninety-six percent of the gorillas in protected, carefully monitored family groups. Spillover of Ebola among closely related gorillas partially explained the sudden, catastrophic die-off, but scientists began to posit the hypothesis that another reservoir host, one more abundant than the placid gorilla, was responsible for the unexpected number of deaths in the ape community. The gravity of concern among animal conservationists for the threat of extinction of the entire species heightened when they added to the Ebola death toll the ongoing scourge of commercial hunting of gorillas.

There was a pressing need to discover the ultimate source of the limited Ebola outbreaks that occurred between 2001 and 2003 in Gabon and the Democratic Republic of Congo. Though gorillas and chimpanzees were some of the earliest victims, there was a widely-held belief that a greater multitude of smaller forest animals served as a reservoir in which the virus could grow to the astronomically high numbers needed to perpetuate its tortuous journey to other subhuman primate species and then to humans. In 2004–5 teams from Gabon, France, and South Africa met in the central African regions that had suffered extensive losses of gorillas and chimpanzees. They captured over a thousand small animals, including birds, bats, and rodents. A number of bats, representing three different species, carried antibody to Ebola virus, proof of prior or current infection. Viral RNA was detected in liver and spleen, but the RNA load in the affected organs was low, which explained why whole infectious virus could not be recovered easily from the bats; the virus was probably there in small amounts. In some bats, viral RNA was present but antibody was not, indicating that they were very recently infected, before antibody was generated.

Each of the bat species found to carry EBoV has a broad geographic range which overlaps regions where Ebola outbreaks have previously been recorded. During the dry season, when fruits become less plentiful, apes and bats come into closer contact as they vie for the same food. And periods of relative hunger, or even starvation, weaken animals and their immune functions, conditions that allow viruses to thrive.

Worries about the potential eradication of magnificent animal species in Africa and concern about the apes' relationship to bats became the backdrop to the impending threats Ebola posed to the human species. In the early twenty-first century the whole world stood at attention and joined hands in a concerted effort to learn more and do more. Between May and November 2007, Ebola infected more than 250 people in the Democratic Republic of Congo (DRC); over 180 victims died. Then, in March 2014, Médecins Sans Frontières was called in to aid local health workers in caring for clusters of patients laid low by vomiting and bloody diarrhea; some had already died from dehydration and shock. Ebola was the apparent cause.

Under the auspices of the WHO, an international team of investigators from Gabon, France, DRC, and Switzerland traced the earliest stages of the 2007 outbreak. They conducted several epidemiologic investigations that included collection of social and environmental data, and did analyses of ecological conditions and animal population surveys. Villagers in the most affected area of the DRC were interviewed. The locals hadn't noticed any increase in the number of illnesses or deaths among their domestic animals or their favorite forest protein sources. But they did report a massive incursion of fruit bats, many thousands more than in any recent years. The bats came in April and May when the fruit trees were heavy with their bounty and settled in the abandoned palm trees of a local plantation. It was claimed that the plantation had probably served as a luscious stopover site for migratory bats every April for at least fifty years. The local hunters felt fortunate; they had endless access to bat flesh, a major source of dietary protein to feed their protein-starved tribespeople.

Over the years, I've learned the value of a careful review of the natural history of a patient's illness, including his family constellation and its pets; his exposures to other animals;

his travel and dietary habits and the medicines he takes; and his contact with other sick persons. All of these contribute to the construction of a tentative differential diagnosis, one that guides the selection of laboratory tests and radiographic procedures. The results of these tests narrow the list of possibilities to those few that focus the work-up still further. It's mandatory that one consider the patient in the context of his community, to think of the patient in all his complexity, but as only one element in his unique ecosphere.

One such systematic investigation of victims of the 2007 DRC Ebola incident revealed that the outbreak began in May 2007 in a colony of ten villages situated 200 km from a notorious 1995 outbreak in Kikwit. The index case was a 55-year-old woman who succumbed to classic Ebola hemorrhagic fever syndrome, complete with vomiting, bloody diarrhea, and hemorrhage into multiple organs. She died on July 3, 2007. Eleven of her contacts, most of whom were members of her family, fell ill five to ten days after she first got sick. They all developed typical viral hemorrhagic fever and died within a few days. In the course of their interviews, the investigators soon learned about the death of a four-year-old girl who got sick on June 12, 2007, and died on June 16 after one long episode of fever, vomiting, and diarrhea. How did she get Ebola? How did the 55-year-old woman get Ebola? Was it possible to find any connection between them or with the fruit bats, who had, by this time, abandoned the region? It turned out the little girl's father got sick just before his daughter did, but his illness was brief and mild. It's very likely that he was exposed to bat blood, as were many of his acquaintances, when he purchased some bats fresh from the hunt. During his illness, the attentive dad carried his child on the road that linked two of the bush villages involved in the outbreak. The child had almost certainly been exposed to her father's sweat or oral secretions, both of which body fluids are now known to carry the Ebola virus. In keeping with local custom, the fifty-five-year-old woman helped the dead girl's mother and grandmother prepare her body for burial. The body was washed on June 17, and the woman fell ill about eight days later, precisely within the time interval that characterizes Ebola's incubation period.

A unique set of circumstances sparked the 2007 epidemic: a colossal bat migration; thousands of trees ripe with tasty fruit; hundreds of skilled hunters armed to the hilt and determined to bring a treasure trove of freshly-killed bats to their protein-starved kinfolk; a father who bought one or more bats for his family to consume; a blood-tainted father who carried his four-year-old daughter along a dusty road to a nearby village; a little girl who died after being exposed to her father's virus-contaminated sweat or saliva; a middle-aged woman who helped her friends wash the body of the dead girl; a whole family and group of acquaintances who cared for the fifty-five-year-old woman as she died in the throes of Ebola hemorrhagic fever, and so on . . . Most enzootic epidemics start this way, with an exquisite particularity of participants and events which, in sequence, light a fuse that starts an epidemic. The amazing specificity and great unlikelihood that a multifactorial sequence such as this will take place should reassure us. It seems unusual that such unique sequences would ever take place. But the fact that animal-triggered epidemics and pandemics have happened more than once in some of our lifetimes has taught us that the odds are sometimes against us. These scenarios stimulate us to seek knowledge and enact measures that will, at the very least, mitigate the catastrophic potential of future epidemics. The discovery that bats are the likeliest mammalian reservoir for Ebola virus provides ammunition for a judicious public health campaign whose recommendations would include the following admonitions: "Stop hunting fruit bats, especially during seasons when dead chimpanzees and gorillas litter the forest floor. Don't bring bloody ape carcasses or freshly killed bats into your homes."

Many viruses persist in their hosts long after the initial infection and its attendant clinical manifestations fade away. Some viruses, like Ebola, find sanctuary in organs that impede the influence of the immune system, like the brain or the reproductive organs; the AIDS virus, HIV, is another such virus. It is precisely this sneaky behavior that has foiled attempts to "cure" the infection, that is, to rid the body of every last infected cell and every last infectious virion. HIV is particularly sinister because it targets the very cells that orchestrate many of the immune system's actions, and these cells, the helper T cells and macrophages in particular, are found in every lymph node in the body and in lymphoid cells throughout the GI tract, the liver, and the spleen.

The herpesviruses pose the same threat. Herpes simplex virus, the agent of cold sores and genital ulcers, remains latent in peripheral sensory nerves and in the agglomerations of neurons that form the spinal cord ganglia. The Epstein-Barr virus, the cause of infectious mononucleosis, remains latent in B cells, elements integral to the function of the immune system, where it resides for the rest of the host's life. The virus regains its infectious potential under conditions of immunosuppression or other unknown external influences. Varicella-zoster virus, the cause of chickenpox, also achieves latency once the pocks are all gone, only to reappear many years later when the immune system slows down and the awful scourge of shingles, or zoster, rears its ugly head in the form of an inflamed vesiculopustular rash in the precise distribution of those sensory nerves that lay just below the original chickenpox rash, mostly on the thorax, but also on the neck or face.

Evidence from the latest Ebola epidemic now confirms that this virus, too, can persist in semen and vaginal fluids long after the symptoms of the disease have subsided. The value of molecular medicine, when put to use as a diagnostic tool, was demonstrated in a straightforward but laborious set of experiments orchestrated by scientists from the Centers for Disease Control and Prevention, the University of Massachusetts Medical School, the New York University School of Medicine, and the Warren Alpert Medical School of Brown University. The researchers detected Ebola virus RNA in the semen of male survivors of Ebola disease in Liberia for as long as 290 days after disease onset. Fully infectious virus was also cultivated for as long as eighty days. The precise genetic signature of the virus was also found in some female partners of infected men. Testicular tissue gets infected during periods when the virus is in the blood, and the testes are believed to be at least partly shielded from the host immune response, allowing the virus to replicate unimpeded for some extended period of time and at least as long as sexual transmission occurs. This is all worrisome, in that one can never know with certainty when any individual infection ceases to be transmissible through sexual contact. But even more significant, health officials have not been able to safely declare that an epidemic is over. We cannot monitor all infected persons for their continuing status as viral shedders, nor can we tell every infected person when he or she can safely have sex again.

As a result of these knowledge gaps, the World Health Organization has recommended with its usual conservatism that male Ebola virus survivors should abstain from sex or practice safe sex using latex condoms until semen has tested negative twice in at least a week-long interval. WHO further advises that for male Ebola survivors whose semen has not been tested, safe sex must be practiced for twelve months following onset of Ebola disease. Who's going to undergo such testing? Who's going to do these tests? Who's going to buy the expensive equipment needed to perform the assays? Who's going to follow these stringent rules? The WHO has promulgated recommendations that have little chance of being implemented in a greatly compromised, resource-limited continent. The prevalence of sexually-transmitted diseases is common in many countries. Are we to recommend that all partners whose current and past STD histories are unknown engage in safer sex practices at all times? Perhaps we can, at least, say the words. Will these admonitions stop the progress of all epidemics that have the potential for sexual spread? Almost certainly not, as the ongoing HIV pandemic has already shown.

Long after EBoV has disappeared from the blood, it sometimes persists in the brain. It has been recovered from the cerebrospinal fluid of Ebola patients with encephalitis, even up to nine months following recovery from the acute disease. Affected patients exhibit behavior characterized by apathy, slow ideation, headache, and aggressiveness. Later they suffer from dizziness, gait instability, disinhibition, and incontinence. The virus, like many others (Chikungunya and rabies, for example) can enter the central nervous system hiding inside phagocytic white blood cells (some refer to this as a "Trojan horse" attack), or by way of cells lining the interior of small-diameter blood vessels. Virus has also been cultured from the eyes of expatriate healthcare workers, long after they've left West Africa. It's not yet known whether infectious virus persists in such "privileged compartments" in small, nearly undetectable numbers beginning from the earliest stages of infection or whether it emerges from a quiescent latent state when provoked, from whence it re-enters the bloodstream. It's been postulated that some of the unproven but potentially life-saving treatments administered to some of the earliest victims of Ebola may have unexpectedly dampened those elements of the immune system needed to keep the virus at bay.

Liberia made two "Ebola-free" declarations before Ebola reappeared in the country. Sierra Leone announced a new outbreak about two months after claiming itself to be free of Ebola; the announcement came just seven hours after the World Health Organization declared the end of all chains of Ebola transmission in Africa. No one knows whether these returns are caused by some new animal-to-human contact, by the re-appearance of previously quiescent infectious virus in an index patient, or by some contact of the putative index patient, perhaps a survivor of the most recent outbreak, with someone who was shedding the virus in the absence of symptoms.

What do we do to guard against these unwelcome recurrences? The tactics that have been proposed are complex, expensive, and difficult to enact. Health professionals suggest that we counsel widely regarding the hazards of unprotected sex and organize programs to follow Ebola survivors long-term so as to monitor the incidence and duration of episodic recrudescent infections. It's been further suggested that local health agents step up surveillance for new cases and encourage scientists to improve the sensitivity and availability of Ebola virus detection assays so that isolation procedures can be put in place quickly. Current estimates indicate that the earliest Ebola victims transmit infection to about thirteen percent of household contacts and that transmission of infection is most likely to occur in those who provide direct nursing care to ailing victims. However, about twenty-seven percent of Ebola infections are asymptomatic and thus unrecognizable—a further reason to make available accurate and rapid diagnostic tests so that all infections are detected, not just those that rise to the level of overt illness. Even those who shed the virus without symptoms are infectious. Diagnostic advances, ready to use at the bedside (point of contact testing), must be fully in place before antivirals and vaccines are introduced during the next epidemic. The efficacy of these treatments cannot be accurately assessed until *all* infections are counted, those that are clinically apparent and those that are not.

The notorious 2013 outbreak of Ebola virus disease in the impoverished, war-torn countries of West Africa gripped the world's attention. There was concern for pandemic spread. The world had grown progressively smaller in the prior four decades. In fact, there had been twenty-nine well-documented outbreaks of Ebola and three of Marburg disease between 1967 and 2014. All told, over twenty-eight hundred people had fallen ill. This number does not include the nearly twenty-eight thousand who succumbed during the 2013–14 epidemic in Guinea, Liberia, and Sierra Leone. In the majority of outbreaks, one-half to three-quarters of patients died. Investigators from the Tulane School of Public Health and Tropical Medicine and Doctors for Global Health in Atlanta carefully analyzed the available data and discovered that in six of the twenty-nine documented outbreaks the infection could be traced to the hunt for and consumption of non-human primates. In six additional instances, exposure to gold mines and their resident bats or exposure to bats through hunting are believed to have been the trigger for the outbreak. Although secondary spread of disease most often takes place in the community or household setting, a significant proportion of transmission occurs in the healthcare setting! Nosocomial or intra-hospital spread of Ebola clearly took place in twelve of the outbreaks. Since the beginning of recorded time, doctors, nurses, aides, laboratory workers, morticians, and others have been secondary victims of plagues. In a few instances, traditional healing practices likely contribute to spread.

Control of epidemics requires coordinated effort, expenditure of large sums of money for supplies and services, and functioning transportation systems that permit timely access to treatment facilities. The collapse of these efforts in times of civil war, insurgency, and other forms of armed conflict foil all forms of health intervention. The enormous population shifts and interminglings that occur during times of unrest are perfect fodder for viruses that see crowds and decay as their best allies. The recent civil tribal wars in the Democratic Republic of Congo, Sudan, Guinea, and Sierra Leone and the ongoing armed conflicts in these same countries set the stage for the arrival of treacherous little actors sporting filmy cloaks, gaily studded with proteinaceous hooks which grabbed onto innocent bystanders, most of whom were poor and malnourished.

During the most recent Ebola epidemic, the entire world was at risk, given how common international travel is. Therefore, why were there so few documented cases outside the "hot zone"? The obsessive collection of data by international health authorities showed that there were only thirty-four cases of Ebola outside West Africa. Oddly, there were no cases reported from other parts of the continent which had been subject to periodic outbreaks going back to the 1960s.

Four cases of Ebola were first diagnosed in the United States. The first, a native of Liberia, traveled to Texas and died in a hospital on October 8, 2014. Two of the nurses who cared for him got sick, but later recovered. The fourth patient, a medical aid worker, served with Doctors Without Borders in Guinea; his diagnosis was confirmed by a lab at CDC. He recovered and was discharged from Bellevue Hospital Center on November 11, 2014. Four hundred and fifty-eight contacts of the four patients were identified and followed for as long as four weeks. There was no secondary spread! The actions that stymied the spread were old-fashioned, some ancient. Quarantine was back in vogue. Efficient identification of cases and the obsessive use of infection control measures, which include the donning of personal protective equipment or "PPE," worked exactly as hoped for. Handwashing, topical microbicides, face masks and gowns, sometimes boots and headgear, proved to be extraordinarily effective barriers to spread. Unproven vaccines and antivirals were hastily invented and a few were tested in the field. Some may have hastened recovery or quieted the symptoms of Ebola, but none was proven to be effective when subjected to stringent scientific analysis. Isolation, soap, masks, gloves, booties, and gowns, the quotidian signatures of every well-behaved healthcare facility, and even space-station headgear saved the world from the ravages of Ebola. How do we solve a problem like Ebola in a more streamlined way? The same way we've dealt with other microbes in recent history—by inventing vaccines and developing pharmacologics. The most recent Ebola epidemic greatly magnified the clarion call for scientists to get to work.

Vaccines come in two varieties—prophylactic and therapeutic. Prophylactic vaccination or immunization is far more common and is designed to stimulate a hardy, long-lasting immune response, one that will continue to protect the host from his first infection and for years after the vaccine is administered. In order to lengthen the duration of the vaccine's potency, we almost always administer booster shots, because we know by experience that the immune response is "boosted" and made more enduring if we stimulate the body's protective potential by giving it multiple opportunities to fortify its troops of antibodies and warrior white blood cells. Therapeutic vaccines are used far less commonly. True to their name, they are a form of therapy once the disease has already taken hold. They are designed to give the body's immune response a little boost, an extra dose, if you will, of the body's most effective armamentarium. A handful of prototypical Ebola virus vaccines of both types have been developed and have been tested for efficacy in animals, mice, guinea pigs, and nonhuman primates. In these experiments the outer envelope of the virus has been employed as the active component of the vaccine, one designed to elicit a neutralizing response.

Those immunologists who are primed for the battle against the human immunodeficiency virus, HIV, and Ebola virus, have also devoted their talents to creating *passive* immunization products, ones that can be made in relatively large amounts *in vitro*. These immunogens are mixtures of monoclonal antibodies, or MAbs, protective molecules that are directed against very specific viral protein antigens, each of which is unique in its three-dimensional structure. Researchers have shown that a mixture of three mouse-derived monoclonal antibodies was effective in treating non-human primates infected with the 2013 strain of West African Ebola virus. Routine methods already exist for extracting these kinds of MAbs from human lymphocyte cell cultures. One such continuous antibody-producing cell line has been created from the white blood cells of a man who survived a 1995 outbreak of Ebola and who continued to have neutralizing antibodies in his blood after twenty years. Unexpectedly, the cells of this long-term survivor produced four different kinds of MAbs; one of them neutralized the virus without the help of the other three. Another MAb, created by scientists in France, was also directed against an Ebola virus outer envelope protein. It blocked the ability of the virus envelope to fuse with the host cell membrane, thereby preventing the virus from ejecting its genetic code into the host cell.

A final monoclonal antibody cocktail, ZMapp, which gained much publicity during the 2014 epidemic, had some limited success. ZMapp was the result of collaboration between investigators at the Public Health Agency of Canada, Kentucky BioProcessing, and the National Institutes of Allergy and Infectious Diseases (NIAID, a branch of the National Institutes of Health). The antibodies comprising ZMapp were designated as chimeras because their components were passed at various points in their genesis through mice, guinea pigs, and primates to find the very best combinations that could ablate replication of Ebola in rhesus macaques.

In 2014 a limited supply of ZMapp was tested in seven individuals infected with Ebola; of these, two died. Soon after, it appeared that supplies of ZMapp had been exhausted and that no further treatments could continue in the absence of data proving the ZMapp was effective and safe. However, soon thereafter more robust production of ZMapp began by way of a contract with Kentucky BioProcessing, a subsidiary of Reynolds American, the tobacco company. Genes coding for the chimeric antibodies were incorporated into viral carriers called vectors. The vectors were then inoculated into tobacco plants, which are able to produce the antibody mixture in huge amounts. Part of the cost of production was supported by the US Department of Defense, which has had a healthy interest in this methodology as part of its enhanced biodefense operations in the wake of the attacks on 9/11/2001.

In March 2015 a large enough supply of ZMapp was made available to justify beginning a randomized, controlled trial in the field. Seventy-two infected patients were enrolled at treatment sites in Liberia, Sierra Leone, Guinea, and the United States. After measuring the viral loads in each study subject, they were divided randomly into two groups. Those in the "study" group received "standard of care" and three intravenous infusions of ZMapp. The "control group" received standard of care plus a relatively new antiviral drug which was believed, but not proved, to have some value in treating viruses like Ebola. Members of the control group were *not* given ZMapp. The study was designed to compare the mortality rates in the two groups at twenty-eight days. The death rate was thirty-seven percent in patients who received standard care alone compared with twenty-two percent in patients who also received ZMapp. To naïve observers like myself, this appeared to be a promising outcome. However, the biostatistical sticklers who analyze reams of data from trials like these apply very strict standards to determine drug efficacy. To quote one conclusion of this study: ". . . although the estimated effect of ZMapp appeared to be beneficial, the result did not meet the pre-specified statistical threshold for efficacy," even though there was a forty percent lower relative risk of death with ZMapp. Adverse reactions were similar in both groups.

Prior to launching this trial there was considerable debate about which among the handful of drugs that showed some preclinical activity and that *could* be tested *should* be tested first. Also, the researchers needed to choose the most desirable way to test a new intravenous compound (ZMapp has to be given IV) in the midst of a colossal public health crisis, while they made sure that the trial rigidly adhered to twenty-first-century standards of informed consent. (This is a formidable challenge in countries where a lot can be lost in translation.) If possible, future field trials of this type would need to take into account some of the criticism and unavoidable complications that came to light in the course of the ZMapp study. First, most of the enrollees were one week or more past the onset of their disease, thus delaying the administration of ZMapp beyond the five-day window in which the drug was found to be most effective in nonhuman primates. Second, the need for an intravenous route of administration and the need for three infusions over the course of a week is an almost unattainable requirement in resource-poor countries. In the 2015 trial I've referenced, a full course of ZMapp was given only to those who survived the first week of the trial. Patients who entered the study with higher viral loads, that is, those who, through no fault of their own were enrolled later than was optimal, were more likely to die sooner than those with lower viral loads. IV equipment is expensive, and too few healthcare workers have the skills required to start and maintain the sterility of intravenous access. During the 2014–15 epidemic in West Africa, the Emergency Ebola Treatment Unit in Sierra Leone was the only one able to provide care at the level of an intensive care unit. Third, although study subjects enrolled in the ZMapp trial were randomized, the study was *not* "double-blinded"; that is, both patients and caregivers knew whether they were or were not assigned to the ZMapp arm of the study. Unblinded or open-label drug trials introduce the possibility of "observational bias" at the bedside. Although this type of bias does not affect the determination of mortality, it may influence a host of other subjective and objective observations and measurements that are crucial to the conclusions researchers draw when the study is over.

Lastly, the study was terminated before the intended number of participants was enrolled. The goal had been to include one hundred subjects in each of the two arms. But there had already been a dramatic decline in the number of infected patients while the study was ongoing! As the Writing Group for the Multi-National PREVAIL II Study Team concluded in their 2016 valedictory:

> The laudable and rapid decline in eligible new cases of EVD was a factor that no trial design could anticipate, and it affected our ability to reach definitive conclusions . . . The outbreak appears to have ended with no incontrovertible evidence that any single treatment intervention, or combination of interventions, was unequivocally superior to the types of supportive medical care typically provided. (*New England Journal of Medicine*, October 13, 2016)

The PREVAIL II team (Partnership for Research on Ebola virus in Liberia II) make a strong plea that future epidemics of Ebola stimulate still more innovative approaches to the disease and that among these some be chosen as "experimental interventions" worthy enough to be evaluated in as "rigorous a manner as possible" in order "that their success or failure can be declared with the confidence that public health policy demands."

We can only hope that potential treatments are in the pipeline, ready to hit the pavement, as soon as the first Ebola cases in some future outbreak are reported. We can also hope that clean water, navigable roads, and storehouses of personal protective equipment are in place so that new treatments can be given efficiently to the people wanting to get well.

It's hard to stop a ravaging epidemic in resource-limited countries once a critical share of the local population has become infected and is contagious. The creation of effective antiviral drugs and preventive vaccines takes a rare form of ingenuity, creative design, and proof of concept. It's unethical to test new treatments without informed consent, and in extreme conditions it's often hard to find the requisite number of study subjects, a number that often ranges between the hundreds and thousands. One must constitute an experimental group large enough to produce exacting numerical results that are attained only after the most complex biostatistical analyses.

Scientists can expect to have the results of their vaccine and drug studies published only if they follow a randomized, double-blind design. That means half the study subjects will receive a placebo or a treatment already proven to be effective, rather than the experimental study drug. It also means that neither those persons dispensing the experimental drug nor those taking it know whether they're being given the drug or the placebo. It's also likely that the results will only be published if they're positive. Some of us object to that stringent requirement, because even negative results may be helpful when designing new studies in similar settings. But that's the way it is. Exacting studies like these are also exceedingly expensive when one adds up the costs of exhausting laboratory research and design, the manufacturing costs, and the tedious field studies—finding vaccinees, doing informed consent in the native language, and assiduously following the patients for days, weeks, or even months following the initial roll-out. (It's especially important to learn whether the new drug produces adverse side effects.) Add to that the numerous laboratory tests that must be done to prove efficacy.

Despite these numerous roadblocks, half a dozen or more vaccine and antiviral drug studies were initiated during the 2014–15 epidemic. A few produced promising results when the data were subjected to preliminary or interim statistical analysis. Unfortunately, there were problems that made it necessary to halt some studies prematurely. The results of other studies were rejected by publishers because of errors in study design or the failure, not intentional, to enroll enough study subjects or of failure to follow the most stringent rules for selecting untreated controls. Many publishers will not publish studies in which "historical controls," selected from previous studies, are chosen as substitutes for "contemporary controls." Some of the studies started too late. A silver lining accompanied the late start; by the first quarter of 2015, containment efforts, quarantine and community engagement, led to a sharp decline in new cases. By that time, it was too late to find enough newly-infected patients or persons at risk of getting infected soon.

One study enrolled a mere ten patients. Another study, designed to evaluate the protection afforded by a GlaxoSmithKline vaccine, needed to enroll nine thousand people to achieve statistical significance; only five hundred volunteers came forward. Note the word "volunteers." International ethical standards governing human experimentation require adherence to very strict rules of informed consent.

Sometimes, even in highly educated, wealthy populations, there is a smidgen of distrust of pharmaceutical companies hawking their new untested wares. (As we all know, even FDA-approved, well-tested products are sometimes rejected by wary patients.)

In Guinea, government regulators entirely rejected a randomized, placebo-controlled trial of a new drug. They believed, some would say rightly, that it was unethical to withhold current standard of care and treatment from the controls. Giving a sugar pill or an infusion of glucose water to the ailing controls seemed terribly wrong to the regulators. On the other hand, drug company scientists argued that any data favoring the efficacy of a new drug would add important information to the quest for a cure or preventative and that a randomized placebo-controlled study design was getting in the way of progress, especially in the midst of a fast-moving fatal disease outbreak. In a study published in *Clinical Infectious Diseases* in June 2016, Lori Dodd and her colleagues at the National Institutes of Health made a plea, supported by complex biostatistical analytics, "that future studies may require acceptance of a paradigm that circumvents, accelerates, or re-orders traditional phases ["of study design"] without losing sight of the traditional benchmarks by which drug candidates must be assessed for activity, safety, and efficacy." Dodd and her co-authors argue for more direct "bridging from preclinical studies to human trials than the conventional paradigm would typically have sanctioned."

Applying that very philosophy, one vaccine study stands apart from the rest: heroic efforts were mounted to get the study off the ground quickly, as conditions on the ground worsened. The methods employed were innovative. As reported in the magazine *Science* in the summer of 2015, the "WHO faced the challenge of a lifetime . . . coordinating a vaccine trial with an untested design, . . . in one of the world's poorest countries, where the health system is dilapidated and the population mistrusts both its own government and foreign assistance." The WHO enlisted the help of the Norwegian Institute of Public Health, Doctors Without Borders, and a dozen other international institutes and labs. The vaccine, a chimera of sorts—a cattle virus, VSV, which does not cause disease in humans—was hybridized with a highly immunogenic protein found on the surface of the Ebola virus. The study design was borrowed from the most successful antiviral battle of all time, the one mounted against smallpox in the latter part of the twentieth century that resulted in the ultimate eradication of smallpox from the face of the Earth. The technique is called "ring vaccination," a process by which the vaccine is administered as quickly as possible to the people most at risk, that is, those reported to be direct contacts of well-documented Ebola virus victims as well as the direct contacts of the primary contacts. In the initial Merck vaccine run-through, ninety clusters of contacts and controls, comprising 4,400 vaccinees, were enrolled and studied.

The British journal *The Lancet* published an interim analysis of the results. Some problems with the study design were revealed: No one knew exactly how long it takes the vaccine to elicit a protective antibody response. Therefore, failure of the vaccine to be deemed effective might simply be due to the extended time needed for it to produce the desired antibody response. No tests were done to identify those vaccinees who were already infected with the wild virus but still asymptomatic at the time of immunization. As a result, it would be impossible to judge whether a protective immune response was due to the wild virus or the vaccine. The Merck vaccine contained an Ebola strain from the Democratic Republic of Congo, but all the vaccinees were infected or exposed to a viral strain isolated in Guinea. It was believed, but not explicitly proven, that these two Ebola virus strains were so similar that they could be expected to elicit cross-protective immune responses. Despite these imperfections, the study's authors concluded that the vaccine efficacy exceeded seventy-five percent, a value that encouraged one study author to state that he was ". . . extremely confident this vaccine is highly effective." Other experts from around the world echoed their enthusiasm, so much so that the trials were re-started in Guinea. Health officials in Sierra Leone and Liberia discussed the possibility of joining the study, even though the vaccine remained unlicensed. Ana Maria Henao-Restrepo, a Colombian epidemiologist and senior author of the study results, made the most prescient remark: "This is really an example of what we can achieve if the international community works together." Soon thereafter, deliberations commenced regarding possible trials with a competitor vaccine, created by Glaxo-Smith Kline.

Another international group of investigators from Belgium, France, Senegal, the Ivory Coast, and Guinea made a case for the kind of study they'd already completed. Their aim was to measure the results of treatment of Ebola patients with favipiravir, a new antiviral, one similar to a drug that's been used for other viral infections. Their data were published in *PLOS Medicine*, a relatively new online journal. Here's what they said (the italics are mine):

> We did a *non-randomized trial*. This trial reaches *nuanced conclusions*. On the one hand, *we do not conclude* on the efficacy of the drug, and our conclusions on tolerance, although encouraging, are *not as firm as they could have been* if we had used randomization. On the other hand, *we learned about how to quickly set up and run an Ebola trial*, in close relationship with the community and non-governmental organizations; we *integrated research into care* so that it improved care; and *we generated knowledge* on EVD that is useful to further research.

I applaud their heroic efforts and marvel at their conclusions, expressed in such fuzzy, "nuanced" prose. Perhaps scientific papers like this will more easily find their way into the public sphere. The perfect should not be the enemy of the good. As educated people in a free society, we should be able to sort through all sorts of information and honor work that's really good, but not necessarily perfect. We learn from other people's imperfections. Great discoveries are made by a process of slow accretion and repetition, almost always gathered over a period of years, if not decades. It's beneficial when scientists and students working in their separate silos learn from one another in real time, even as they continue to re-think and modify their experimental designs and methodologies, based on the less-than-perfect work done by others.

CHAPTER 9

A Witchy Brew of Bat Spats, Horse Froth, and Womb Drips

NIPAH AND HENDRA VIRUSES

> Outbreak control in Malaysia has focused on culling pigs in the states of Perak, Negri Sembilan, and Selangor; approximately 890,000 pigs have been killed. Other measures include a ban on transporting pigs within the country, education about contact with pigs, use of personal protective equipment among persons exposed to pigs . . .
>
> *Morbidity and Mortality Weekly Report*, April 30, 1999

Four outbreaks of Nipah virus disease were reported from various parts of Bangladesh during the years 2001 to 2004. They all occurred between January and May. In one outbreak, disease often followed contact with sick cows, in the next outbreak with a herd of pigs and, in 2004, with the trees in which bats had nested the night before. To avoid a fifth outbreak, the Institute for Epidemiology, Disease Control and Research of the government of Bangladesh mounted an investigation and invited the Centre for Health and Population Research to assist. A team of anthropologists soon joined the investigation. Their stated goal was to identify risk factors for acquiring Nipah infection, with the ultimate plan to design ways to prevent it. The investigators were particularly troubled by the rising incidence of encephalopathy and seizures in those afflicted.

A classic case-control study design was chosen. The criteria for selecting "cases" were strictly defined. The "controls" were persons close in age to the cases and living in the next closest house to a case. The anthropologists designed detailed questionnaires that were administered to both cases and controls. For case patients who were unconscious or had died, surrogate respondents were selected who knew a lot about the patients' daily habits and terminal illness. As per standard procedure, there were three times as many controls selected as there were cases. All study participants had blood drawn for virus antibody analysis.

The median age of case patients was sixteen. Most were male, and ninety-two percent had died a short time after an illness that began with fever and unconsciousness. There was only one exposure and habit associated with the disease that was statistically significant: consumption of raw date-palm juice. Those who had drunk the juice were eight times more likely to succumb to Nipah virus disease than those who had not. A few of those who fell ill had been in contact with another Nipah patient.

Fatal viral disease outbreaks never originate from a wild botanical source. People catch viruses from other people or from insects that bite them and carry viruses they've acquired after biting infected animals or people. And, as I've shown in this book, people sometimes catch viruses firsthand from the bites, blood, secretions, or excreta of animals, almost always warm-blooded ones, like mammals or birds. The virus hunters in Bangladesh were obliged to find the link between Nipah virus disease and date-palm juice. Some of them went into the forest to learn how this apparently deadly liquid was collected.

PBS produced a film called *Spillover: Zika, Ebola and Beyond* that aired over a number of their affiliated TV stations in the summer of 2016. The show provided current information about the ongoing Zika virus epidemic in South America and Caribbean. It also produced segments about the disease outbreaks caused by the Ebola and Nipah viruses to remind its audience about the ominous relationship between animal infections and humans. The segment on Nipah was astonishing. The cinematography was gorgeous. It evoked in me a jaw-dropping response. I remember gasping before I uttered "you have got to be kidding!" It turns out that date-palm juice, really sap, is a Bangladeshi delicacy—people love it, and so do fruit bats. The sap runs best during the early months of the year, January through March. Agile young men grab onto the palms with both legs and arms and climb the tree in a kind of galloping, leaping motion. While dozens of feet off the ground, they cut a tap into the trunk and hang a clay collecting pot. Bats of the genus Pteropus also pierce the palm tree trunk with their teeth and lap up tiny amounts of sap. However, those few drops are nothing compared to the riches that gather in the hanging pots. While making their own little slashes in the trunk, the bats sometimes urinate and defecate into the pots hanging just below. At other times they cannot resist the temptation to cling to the edge of the pots and drink directly from the taps or the vessels themselves. Of course they cannot help dribbling droplets of saliva into the pots or peeing on their edges. The date juice is greatly sought after and quickly sells out. Connoisseurs know that if they wait too long the juice will begin to ferment, marring the sap's natural sweet flavor. Fresh is best! The virus thinks so, too.

Nipah virus spreads indirectly from bats to humans, but it also spreads directly from person to person. Detailed investigations of outbreaks in Bangladesh, Malaysia, and the West Bengal region of India regarding the incidence of interpersonal spread of Nipah documented the risk factors associated with transmission. Two such investigations, one undertaken in Faridpur, Bangladesh, in 2004 and the other, the result of a retrospective study of an outbreak in Siliguri, India, in 2001, proved that family members and hospital workers were at greatest risk. Nipah virus is shed in high titers in respiratory secretions and urine. In the Faridpur outbreak, which ultimately fanned out through thirty-four people in five successive cycles of transmission, the original cluster comprised five individuals, the index patient, three family members, and a neighbor who cared for him. The caregivers got sick within two to four weeks of the onset of the first victim's illness. The index patient's aunt fell ill, and a popular religious figure cared for her. When he succumbed to the illness, he was visited by numerous individuals in his community; twenty-two of his devotees developed Nipah virus infection. Among these, one returned to his home village. The rickshaw driver who transported him and a few of his relatives back home got sick.

Direct physical contact with a Nipah patient proved to be the strongest risk factor for acquiring infection. Investigators from the International Centre for Diarrheal Diseases Research in Dhaka and from the CDC in Atlanta teamed up with anthropologists. Their observations were summarized in a 2009 issue of the journal *Clinical Infectious Diseases.*

> [There were] multiple opportunities for the transfer of NiV contaminated saliva from a sick patient to care providers. Social norms in Bangladesh require family members to maintain close physical contact during illness. The more severe the illness, the more hands-on care is expected. Family members and friends without formal health care or infection control training provided nearly all the hands-on care to Nipah patients both at home and in the hospital. Care providers during the Faridpur outbreak continued to share eating utensils and drinking glasses with sick patients. Leftover food offered to ill Nipah patients was com-

> monly distributed to other family members. Family members maintained their regular sleeping arrangements, which often involved sleeping in the same bed with a sick, coughing Nipah patient. There was a particularly strong desire to have close physical contact near the time of death, demonstrated by such behaviors as cradling the patient's head on the family member's lap, attempting to give liquids to the patient with a spoon or glass between bouts of coughing, or hugging and kissing the sick patient.

The consequences of intrahospital spread were documented in an investigative report of a Nipah outbreak affecting sixty-six people in Siliguri, India in 2001. Hospital personnel who work in countries that have no infection control measures in place or that have such protocols but fail to enforce them are constantly exposed to blood or body fluids that contaminate bare skin or splash into eyes, noses, and mouths. Needle sticks may occur during routine blood draws or placement of peripheral intravenous lines or deep vessel catheters.

The Siliguri outbreak started in a hospital and spread to four others. Eventually it involved forty-five hospitalized individuals and twenty-one contacts. Of the patients admitted to the hospital, all had a diagnosis of encephalitis, characterized by fever and often headaches, myalgia, confusion, and involuntary movements. About half were breathing fast, to the point of distress. Most of the victims had lost their deep tendon reflexes. All of them died within a month of hospital admission.

Before the outbreak, hospital workers only occasionally adhered to standard infection-control protocols. There was minimal use of gloves and masks. Hand washing was often neglected, except in the ICU. Waste, soiled linen, and used clothing were handled haphazardly. There were no rules in place for visitors. When it became clear that an encephalitis outbreak was really happening, stricter infection control measures were instituted.

No one knew which infectious agent was responsible for the outbreak, so diagnostic tests were performed for all the usual suspects: Japanese encephalitis, West Nile, dengue, measles, and leptospirosis. Not surprisingly, almost all the patients had antibodies to measles virus; most had either been vaccinated for measles or had had natural infection. Tests for all the other agents were negative.

Siliguri is located close to the Bangladesh border, and when the doctors remembered hearing about Nipah virus outbreaks in their neighbor to the east, they tested blood and urine samples for Nipah. Half the blood samples contained both acute and convalescent antibodies to Nipah virus, and viral RNA was detected by PCR in four urine samples from antibody-positive patients. When subjected to gene sequence analysis, there was ninety-nine percent homology between the Nipah viruses in Siliguri and those in nearby Bangladesh and slightly less homology with Nipah virus isolates from Malaysia. This conforms with the belief that Nipah virus, like many other enzootic viruses, evolves extensively in the local animal reservoir before it's later shared with humans. No attempt was made to discover either a direct or indirect animal contact in the Siliguri outbreak. However, the presence of abundant numbers of *Pteropus giganteus* fruit bats in Northern India now drives the worry that Nipah virus disease will become an ongoing threat.

Bats of the genus *Pteropus* are better known as flying foxes. Their geographic range is enormous. They inhabit Madagascar and Mauritius in the western Indian Ocean as well as New Guinea, Indonesia, and the Cook Islands. Some also live just south of the Himalayas. They range in weight from approximately three-quarters of a pound to a little more than two. Their wingspan is the largest in the world for bats. In some species the wings are a bit over five feet from tip to tip. Their senses of smell and eyesight are keen, and they navigate successfully at night. They love to eat fruits and flowers, and they roost together in treetops.

Flying foxes are prodigious travelers. Radio-tagged animals have been tracked over distances of 360 miles. The many species overlap in their journeys, they mix together in their roosting camps, and they're in constant search of the best blooms on the most fertile trees. There appears to be no particular regularity to their movements and no evidence that individual species seek companionship with one another as they migrate from season to season or year to year. It's one big game of mix and match, all in the service of a good, fruity floral feast. This behavior creates the perfect storm for viral spread, admixture, and an infinity of opportunities for genes to swap pieces with one another and to come up with new ways to invade and cause trouble for new hosts.

Bats and pigs have long been partners in their shared habit of spreading pathogenic viruses to humans, both directly and indirectly. Unlike the bat-driven Nipah outbreaks in Bangladesh and India, those in Singapore and Malaysia were traced directly back to pigs. Malaysia suffered a large outbreak of febrile encephalitis disease in March and April 1999. Two hundred and fifty-seven cases were reported to the Ministry of Health. A hundred patients died. By mid-April the outbreak all but stopped. Epidemiologists learned that over eighty percent of patients had handled or merely touched pigs close to the time of illness onset. Fatalities were particularly high among abattoir workers. Two-thirds of patients reported that the pigs with which they had contact appeared to be ill. It turns out that the half-eaten fruit that falls from the mouths of sated bats is often used to supplement the slop that's fed to the pigs.

Clinicians in Malaysia knew that Nipah virus was very contagious, but investigations of the transmission pathways in various healthcare settings revealed no instances of nosocomial spread, not even to the pathologists who performed post-mortem exams on the deceased. It's been hypothesized that some strains of Nipah virus, like those in Bangladesh but unlike those in Malaysia, are more readily transmitted due to their genetic make-up. Perhaps the more virulent Nipah virus subspecies are endowed with envelope hooks that can attach to the host cell membrane with greater avidity. Or perhaps these Nipah subspecies replicate more efficiently in host cells and, as a result, are shed in numbers so great that they're more likely to reach a new host cell in their vicinity, simply by chance alone.

It's likely that isolation and infection control practices were taught and enforced more routinely in Malaysia than in Bangladesh. As a wealthier country, Malaysia probably had greater access to personal protective equipment (PPE), which, when used properly during a large outbreak of infectious disease, can be quite costly. The fastidious use of the finest PPE available during the recent epidemic of Ebola virus disease in West Africa hastened the successful control of the disease and prevented its spread to other parts of Africa and the rest of the world. The savings attributed to the routine use of PPE in all healthcare settings are incalculable.

The control of Nipah virus disease in Malaysia was ultimately traced to the eradication of infection among pigs. Approximately 890,000 pigs were killed, and the transport of pigs from place to place was banned. However, some small concern was raised when tissues removed from a routine canine necropsy were incidentally tested for Nipah virus; the results came back positive.

Nipah virus shares the same modus operandi on its path of destruction with the few other members of the *Henipavirus* group. They infect and inflame blood vessels throughout the body and exhibit a particular penchant for cells in the central nervous system, including the spinal cord. Thrombosis of blood vessels is a byproduct of the death and slough of the cells lining the vessels. The henipaviruses, of which group Nipah is a member, are unusual in their tendency to induce syncytia: The membranes between cells lining medium and small blood vessels break down in the process of fusion, so once-separate cells become giant cells that harbor multiple nuclei and other organelles. The nuclei become such perfect, fertile ground for viral replication that they swell as they fill with viral proteins that squash all the chromosomal elements to the inner edges of the nuclear membrane. Syncytium formation is also a characteristic of members of the paramyxovirus family, such as measles virus, respiratory syncytial virus, and some of the parainfluenza viruses. In southeast Asia, Nipah virus encephalitis has often been confused with Japanese encephalitis (JE), which is prevalent in the area. But classically-trained, old-school neuropathologists who examine the stained brain sections of encephalitis victims know that JE virus, unlike Nipah, never induces giant cell formation.

Like measles virus, a distant cousin, wild Nipah virus enters the human host by way of the upper respiratory and oropharyngeal epithelium. All primary replication events occur there, and viral progeny travel just a short distance to the abundant lymphoid tissues nearby. There, replication goes gung-ho and viral babies become so numerous that they can fill the bloodstream with enough offspring to infect organs throughout the body. In the brain the virus occasionally goes dormant, from which state it can recrudesce and take the form of a serious acute or more lingering subacute encephalitis for up to a year following the initial infection.

The henipavirus genus is exclusive. Entrée has been given to only three species: Nipah, Hendra, and Cedar, a newly discovered member of the group which, thus far, seems to be nonpathogenic. Nipah virus captured my attention first because of its very odd connection with bats, pigs, and palm juice. Its geographic limitations to India, Bangladesh, Malaysia, and the Philippines caused me to wonder why it hadn't spread to other nearby countries which share many of the same plants, animals, and cultural habits. Hendra virus displays even more geographical restraint; cases have only been reported from the upper east coast of Australia. Horses, not pigs, are major actors in their spread.

As with other viruses that I describe in this book, bats serve as the reservoir for the henipaviruses. The infected bats don't get sick but they do develop antibody which appears not to block their shedding virus in urine, saliva, and secretions of the uterine cervix that sometimes drip to the ground. The henipaviruses can cross the placenta, so virus has been isolated from the fetuses of a few bat varieties. Whether newborn bats or nursing baby bats can also spread infection has not been studied. In order to find out, it would be necessary to abscond with voracious bat infants and use them as experimental laboratory subjects. Even the most curious scientists have been unable to tear nursing baby bats away from their mother's teats.

In the fall of 1994, several dozen thoroughbred horses at a training facility in Brisbane, Australia, succumbed suddenly to a serious respiratory ailment. Fear spread throughout the racing world, and the entire industry was shut down. Thirteen horses died soon after the illness struck. Quarantine measures were initiated, the number of deaths plummeted, and epidemiologists quickly identified the guilty pathogen. It was originally named equine morbillivirus (morbilli was an older name for measles), but was later changed to Hendra virus (HeV) to identify the Brisbane suburb where the disease first struck. Sadly, the first animal to fall ill was a pregnant mare. To provide her with more regulated care, she was moved from her favorite pasture to a training stable. The old mare died after a mere two-day illness, but not before spreading her infection to a dozen additional horses, all of whom died in the next two weeks. Beyond the more usual signs of all lower respiratory tract infections, the sick horses' faces swelled and abundant, blood-tinged frothy secretions flowed from their nostrils. The animals' distress was striking, and seven had to be euthanized. A few horses proved to be infected but never developed clinically obvious disease. These are the animals we most worry about. Because they never appear sick, we may fail to separate them from their stablemates, even though they shed virus that's transmissible. One horse trainer and his stable hand developed an influenza-like illness; both took turns nursing a sick mare in her last days. They became sick before all the other horses in the stable showed signs of disease. The trainer died in the hospital after his lungs and kidneys shut down. Hendra virus was isolated from sick horses and humans alike.

In the wake of the Brisbane outbreak, samples collected from horses in a previous disease cluster were tested retrospectively. This earlier outbreak occurred in August 1994 in central Queensland, about 600 miles north of Brisbane. A pregnant thoroughbred developed pronounced swelling of her cheeks that extended to the tissues around her eyes. A colt in an adjoining paddock, who had frequent contact with the mare by way of openings in their dividing fence, died eleven days after his stablemate. His illness was a bit different; in addition to respiratory problems and bloody nasal discharge, the colt paced aimlessly and had lots of muscle trembles. One more outbreak of HeV infection was recorded in Cairns, North Queensland, in 1999. Again, the fatal case was an older former racing horse who was euthanized. Copious amounts of frothy yellow secretions were found around her nose at the time of death.

Homogenates of spleen and lung tissue from the dead animals caused extensive cytopathology in tissue cultures derived from African green monkeys. The virus spread by way of syncytium formation: the virus caused adjacent cells to fuse allowing movement of HeV from cell to cell to cell, whilst creating new progeny virus in the giant cells that were created. Although Hendra virus disease manifests primarily in the lungs and, less commonly, in the brain, the widespread nature of the infection was documented when virus was also isolated from liver, kidney, and lymph nodes. In the lungs, the alveolar air sacs were filled with proteinaceous edema fluid. Other alveoli were necrotic, as were the walls of adjacent capillaries, allowing their bloody contents to seep into surrounding tissue spaces. In the brain, many neurons die, and the killing fields are occupied by inflammatory white blood cells and cells whose job it is to clean up the detritus. Hendra and Nipah viruses are different from a lot of others in their propensity to favor vascular endothelium, the innermost flat cells lining all blood vessels. Unlike other viruses, the henipaviruses also come in all sorts of shapes and sizes. Some viral particles are up to fifteen times larger than others, but all the henipaviruses display a delicate double fringe of projections—whose function has not been defined—that decorate their outer envelope.

To date, we continue to wonder whether the henipaviruses will find new hosts to parasitize. But we do know that other mammalian species are susceptible to infection under laboratory conditions. These include cats and guinea pigs. Chances are the right ecological balance of environmental features do not yet exist to allow the spread of Hendra and Nipah to feral cats and guinea pigs. One obscure means of transspecies transmission was mentioned in one of the articles I read. There was an unpublished sighting of spread of Nipah virus to people who were present at the farrowing of infected sows and to the dogs who ate the afterbirth.

The zoonotic infections that we worry about most enter the human arena and thereafter spread readily from person to person. The henipaviruses have never reached that level of concern. Nipah virus spreads like wildfire among pigs in closed environments and among patients and their hospital caregivers. Hendra does the same among horses. But once the infection jumps from bats, pigs, and horses to humans, the chain of infection declines quickly. This virus has not developed morphologic and behavioral features, as a result of cycles of mutation, that result in easy and continuous transit from person to person, even by the respiratory route, which is a virus's preferred transit mode.

There is a certain chaos in the universe, also on our planet. So it's not surprising that despite their many attributes, including superior intelligence and sticktoitiveness, talented epidemiologists and clinical virologists may not foresee the next novel zoonotic outbreak before the first victims fall ill. They may not be able to prevent those first few instances of animal-to-human and human-to-human transmission. As in all complex biological systems, there are simply too many variables, all of which must come together in a very particular sequence in a specific place and time to start the ball rolling. The twenty-five to thirty thousand protein-coding genes in each of our genomes are susceptible to change, making our individual selves more or less liable to fall prey to a universe of viruses that also change in random fashion, sometimes in a self-destructive way, but at other times to their distinct advantage.

As we know, climate change and accompanying environmental change are in a state of constant flux. Ecological change surely affected the unanticipated ups and downs of Hendra virus outbreaks. The catastrophic spillover from fruit bats to racehorses in 1994 set the stage. In the subsequent sixteen years, a smattering of fourteen equine outbreaks, mostly in Queensland, Australia, resulted in four human deaths and heralded a three-month-long cluster of eighteen outbreaks, all in 2011. The geographic distribution also had grown, but why? Hendra virus disease had spread from the north, in Cairns, nine-hundred miles to the south, to northern New

South Wales. To everyone's surprise, one horse succumbed to Hendra in Chinchilla, one hundred twenty miles to the west on the far side of a mountain range. Most of the cases involved one or two horses that grazed on peaceful rural properties, but one incident affected five horses and two staff in a veterinary practice. Of the handful of human cases, all were traced to sick horses. None involved contact with a flying fox or with another person with clinically apparent disease. The temporal distribution of cases was just as mysterious. In the twenty years prior to the banner year, 2011, there were two years with one outbreak, then nine years with two; nine years had none. Where did all the infected bats go in those nine pristine years? Was there a sudden surge in protective herd immunity among horses in eastern Australia? Did horses have inapparent subclinical infections? Did the bean counters suddenly lose track of what was going on?

Neutralizing antibodies to Hendra virus have been found in multiple fruit bat species and in frozen samples of blood taken from horses, even before the very first 1994 Australian outbreak. We don't know how long Hendra virus has circulated in a subdued state among bat and equine populations. We do know that the virus is easily detected in urine collected from ground samples taken from populous flying fox roosting spots. We do know for sure that the surge of Hendra cases in 2011 could not be attributed to a more virulent or more transmissible strain. Careful genome sequencing proved that the 2011 strain was consistent with the genetics of previous strains from nearby locations.

In a paper published in the journal *Current Topics in Microbiology and Immunology*, scientists from the Australian Animal Health Laboratory raised a little red flag when they mentioned that there is laboratory evidence to support the potential for Hendra virus infection in pigs, cats, hamsters, and ferrets. A two-year old Kelpie dog from a property harboring equine cases of Hendra infection developed antibodies to the virus, but did not fall ill. In order to prevent spread of infection to other small animals, the dog was euthanized. Dogs and other small animals, particularly rodents, hang around barns and barnyards, raising the specter that these roaming bystanders, even while asymptomatic, could be carriers and purveyors of Hendra infection. They may be the creatures who harbor the virus in those years between reported outbreaks.

These same authors drew attention to the fact that there were no human cases of Hendra disease in the 2011 outbreak. Why? Was there one tiny viral mutation that significantly reduced the chance of transmission to humans? Or was knowledge of the devastating consequences of Hendra infection so widespread and so well-entrenched in the minds of stablehands and veterinarians that they implemented effective infection control measures, including the use of personal protective gear?

To quote the Australian scientists directly, here are some of their ideas to explain the unusual events of 2011:

> . . . we have hypothesized host factors such as physiological and immunological status, and environmental factors such as food resource availability and climatic variables as plausibly supporting an increased level of infection in flying foxes, the prolonged survival of virus in the environment and the increased likelihood of equine exposure and infection . . . it was observed that additional factors such as pasture state and horse behavior may also play a role in the probability of flying fox to horse transmission . . .

How can we expect public health authorities such as the CDC and the WHO, even with their armies of biostatisticians and supercomputers, to make sense of such an impossible complex of animal, viral, human, climatic, and environmental variables, most of which lie beyond our ken, in the realm of things we don't even know we don't know?

CONCLUDING REMARKS

Local and international engagement, substantial financial investment in ongoing surveillance, and infrastructure enhancement are the holy trinity of the earliest stages of zoonotic disease control. The alert reader may wonder why I chose not to include another three—basic science research, vaccine and antiviral drug development, and highfalutin medical equipment—among the supreme triumvirate of necessities required to quell the next epidemic before it spins completely out of control. To be clear, the latter three are critical, but only to future attempts to lessen the spread of Ebola and similarly grave zoonotic threats. As I hope I've made clear in my essays, we already know which interventions work to suppress the tragic events associated with the initial phases of a zoonotic outbreak, and we know how to make them operational! With the exception of pandemic influenza, we've managed to stop the spread of a handful of zoonoses in the absence of effective antivirals and vaccines. Conversely, we continue to suffer epidemic influenza every year despite our annual roll-out of flu vaccines and antivirals. Some zoonoses have simply petered out on their own for some combination of reasons that remains mysterious. Perhaps the virus lost its potency after passing through a number of susceptible hosts;

there is some evidence that this phenomenon holds true for Ebola. Perhaps, after an initial surge, there were simply too few susceptible hosts left to keep the epidemic going, the result of some unknown but potent endogenous immune capacity that lay dormant in the populace, or some portion thereof.

Ebola virus disease faded out of sight, not because of the ZMapp monoclonal cocktail and not because effective experimental vaccines made timely appearances at the peak of the epidemic. In fact, the most recent pharmalogic approaches to battling newly-emerged zoonotic viruses have fallen short. It may take a decade of basic laboratory benchwork, billions of dollars, and complex clinical field trials before a new drug finds its way into the hands of doctors and nurses. In the 1960s and 1970s marvelous antiviral drugs for influenza and the herpesviruses were licensed, but since then antiviral therapies have been approved for fewer than a half dozen, the largest plurality of which were created to wage successful wars against the AIDS virus and hepatitis C. The principal problem that the zoonoses pose to scientists is their unexpected appearance, seemingly coming out of nowhere. Their ecological backgrounds are so specific and their geography so foreign that, so far, there has been simply no effort great enough to meet the enemy in the brief time needed to stop the spread of disease, not only from animal to human, but from human to human. Vaccines for emerging viral combatants face the same barriers; none has made it to prime time (except for annual iterations of the flu vaccine). There are no FDA-approved drugs or vaccines for Ebola virus, dengue, Zika virus, chikungunya, the SARS or MERS coronaviruses, or West Nile virus. Even if scientists were able to create an effective antiviral for each one, we would almost certainly run into the problem of drug resistance. Viruses, like bacteria—methicillin-resistant Staph aureus (MRSA) and multidrug resistant tuberculosis, for example—carry within their genetic make-up an automatic capacity to develop resistance to anti-infectives, particularly when they're used singly in the form of "monotherapy." There are a few exceptions; some antivirals are broad-spectrum, such as ribavirin, a drug that interferes more widely with certain steps in viral DNA or RNA synthesis, among other of its actions. Interferon is an immune agent which acts against a wide array of animal viruses, but has been shown to have only limited success in clinical practice. Interferon has to be given by injection, and its use for Hep C treatment was associated with many untoward side effects that caused most patients to reject treatment after a few tries. It's also very expensive.

Most recently, pharmacologists have been trying to create structural variations of antivirals already in use, particularly those that have proven their excellence in treating HIV and some of the herpesviruses. Some in this class have been shown to be effective in vitro and in animal experiments by subduing infections caused by some influenza viruses, Ebola, and papillomavirus. A few are just entering Phase I human trials.

There is one new approach: to interfere with those host cell molecular pathways that are most often hijacked by viruses in constructing their own selves from scratch. Some of the host cell molecules targeted are used by viruses to fold proteins that will be incorporated into the viral envelope. Other drugs target pathways that are used by viruses to travel around the cell's interior as they gather the ingredients they need to successfully assemble themselves. The greatest limit to this approach is that the host cell utilizes for its own benefit many of the same molecular pathways as the virus. So pharmacologists will have to determine just how many cellular mechanics can be manipulated without doing irreparable harm to the host while still exerting a fatal blow to the virus. One must always balance minimal cellular toxicity against significant viral lethality.

Ebola was all but eradicated as a result of some of the most time-tested and effective practices of infection control medicine: finding symptomatic patients; tracing all their contacts, especially those who provided direct care; transporting patients to specially designed treatment centers that were able to provide fluid resuscitation, blood products, and isolation; and encouraging families to provide safe burials. Amazingly, with great generosity the international community and global charities donated ample supplies of personal protective gear, including tight-fitting helmets, expensive gloves, and boots. Everyone was taught to use the proper maneuvers for donning and doffing the yellow suits, not a pleasant enterprise in equatorial climes. Diagnostic assays and basic lab equipment increased the likelihood of making accurate diagnoses even in persons who did not already display signs of full-blown diseases. Local communities were engaged, and every effort was made to minimize stigma and the natural inclination to shun those thought to be polluted. The afflicted were urged not to reject the "outsiders" who had come to help.

When the counting had stopped, nearly twenty-nine thousand suspected, probable, and confirmed cases of Ebola were reported; the number included over eleven thousand deaths. Only thirty-six cases emerged from a handful of countries outside the hot zone; all were imported. In Lagos, Nigeria, a city of 20 million in the most populous country in Africa, there were only fifteen confirmed cases. That's it! Nigeria dodged a catastrophe through meticulous case finding and quarantine. Nothing fancy, but it worked, as it has for centuries. Sometimes the old ways are best. West Africa was the case that proved the truth of the maxim.

Every school child learns how to stop the spread of the flu: wash your hands (often!) and cough and sneeze into your elbow—primitive, but effective. But the question remains: How do we forestall the appearance of the next epidemic?

And if prevention fails, how do we put into operation effective procedures to slow the progress of disease transmission once it begins?

Experts like Jan Semenza of the European Center for Disease Prevention and Control and William "Billy" Karesh of the EcoHealth Alliance have focused their careers on devising strategies to neutralize the threat of emerging zoonotic diseases caused by viruses and other microorganisms. The United States Agency for International Development has initiated a project aptly named "Predict," a targeted surveillance operation designed to identify emerging pandemic threats. Mammalian and avian diversity and human population density pair to create toxic combinations. Risky behaviors lead to zoonotic spread—hunting, maintaining village and international meat markets, and illegal trade in wildlife all play a role. Environmental degradation and deforestation bring humans, animals, and the microbes they carry into close proximity. Insect vector-borne disease, like West Nile, yellow fever, dengue, and chikungunya cause millions of deaths each year. Mosquitoes, ticks, and sandflies don't care whom they bite—they just want blood; any mammal or bird will do. But thus far, stories of big "bugs" (not viruses) and insects have not yet made transit from my pen to writing pad.

We are all subject to the quiddities of fate. We are all servants to our bodily needs and psychic siren calls. A hunter enters the nearby tropical forest to forage for fresh sources of protein for his family and fellow villagers. He kills a gorilla or a couple of monkeys or bats. He has no way of knowing which, if any, are infected with monkeypox or rabies, Ebola or Nipah virus. A stable worker in Brisbane, Australia enters the nearby pasture to examine a sick thoroughbred mare. He clears her frothy nasal discharge. The man hasn't heard about Hendra virus, so he moves the mare to a larger stable housing two dozen thoroughbreds. He, some of the horses, and the horses' trainer die of respiratory and renal failure in two weeks' time.

Prairie dog fanciers in America's heartland order some pets from an importer who recently received a batch of giant rats from Africa. He mixes the two groups of rodents—prairie dogs and African rats—in close quarters; he didn't know the Gambian giant rats were infected with monkeypox. (He'd never heard of monkeypox.) He didn't know that prairie dogs could catch viruses from Gambian rats. Three members of a Wisconsin family that welcomes the infected prairie dogs into their home fall ill within twenty days. A group of weary hikers return from a trek to Yosemite falls; they crave sleep. They bed down on damp cots in a damp tent in Curry Village, but only after they clear the cabin's nooks and crannies of cobwebs and mouse droppings. They're already pretty grubby, so there's no need to wash their hands. Several weeks later they learn that ten campers with hantavirus infection had also stayed at Curry Village in Yosemite; three had already died.

Should we stop antiviral drug discovery and vaccine development? No! Outbreaks of animal-borne infection will continue to plague us. Some diseases may be those we've previously encountered, like the ones I've described. Some are more deadly than others and have a natural tendency to spread more widely. Those are the ones most in need of our attention, the ones that scientists and pharmaceutical companies are focused on. It takes years and piles of money to get a new drug or vaccine from "the bench to bedside." It would be great to have an effective Ebola vaccine ready for the next outbreak, one caused by a viral clone that's almost certainly maneuvering through its host animals in a tropical forest right now.

Only with effective surveillance and communication highways will the first cases of some future outbreak be brought to the attention of the responsible public health authorities quickly. A licensed vaccine, stored somewhere in large batches, would be transported to the hot zone posthaste and, beginning with those closest to the index patient, the prophylactic vaccine would be administered in ever-widening rings of contact, as was done in the final siege against smallpox, until everyone judged to be at risk has been immunized. The vaccine would be of little use in the absence of near perfect surveillance for local cases and the immediate notification of regional and then foreign public health authorities. The frozen vaccine would need to be loaded quickly onto airplanes and trucks, and the vaccine would need to be maintained at the proper temperature even in countries that lie close to the equator. The roads leading to the villages and towns at risk would need to be passable, even during the wet season. The trucks and planes would need to be in good repair. The local healthcare workers would need to know how to give shots. There would need to be impenetrable receptacles for disposing the bloody bandages, the used needles and syringes. (Ebola patients would almost certainly carry other bloodborne pathogens, like HIV and hepatitis viruses.)

An effective antiviral drug directed against any of the viral infections I've described would also be welcome. It need not "cure," that is, completely eradicate the very last virus. None of the antivirals currently on the market "cures" the infection, but they're all expected to reduce the number of infectious particles so drastically as to result in the relief of the patient's symptoms and to allow the host immune system to pick up many of the remaining pieces. Residual HIV and the herpesviruses remain latent in their hosts for a lifetime, but most of the signs and symptoms of those infections subside.

Financial investment headed the list of my holy trinity of essential capacities. It costs a fortune to bring a new drug or vaccine to market. It costs less to build treatment centers and fill them with the numerous requisite supplies, but poor countries don't have the resources to purchase these items. When the international community pitches in, it serves to limit the outbreak in the home country but also helps to prevent the spread of disease to a broader mix of victims in faraway places. Influenza, SARS, and monkeypox sped from one hemisphere to the other in short order.

Victor Dzau and Peter Sands, representatives from the National Academy of Medicine Commission on Global Risk Framework for the Future, argue that "pandemic preparedness and response should be treated as an essential tenet of human and economic security" and that "civil societies [should] hold governments accountable and . . . sharpen debates about domestic fiscal priorities." Health economists have calculated that losses from potential future pandemics will range from sixty billion to half a trillion USD per year. As dismal realists, these economists remind us that governments might be loath to put aside the necessary three to five billion dollars a year to upgrade national public health systems because, as a species, we "hesitate to invest in preventing and preparing for something that may not happen." We now know that every once in a while something new does happen, Zika virus infection, for example.

Before we have in hand the next cure or the next vaccine for the next plague, we should, as the Family of Humankind, do what works. We're obliged to commit funds to disease surveillance, identification of index cases and their close contacts. We must separate and quarantine the symptomatic patients, build and amply supply basic health facilities, and train a new generation of healthcare workers to apply those basic techniques and procedures needed by people with serious communicable diseases.

We have to prepare wisely and frugally: The Fates are stirring a cauldron of zoonotic pathogens that we may already know about, but which have not yet broken loose in numbers likely to spark an epidemic and which do not yet carry genetic baggage consistent with ease of person-to-person transmission. Lymphocytic choriomeningitis virus, LCM, causes meningoencephalitis and congenital defects in the offspring of infected, pregnant women. It is spread from house mice to humans by inhalation of contaminated aerosols of urine and fecal droppings. Thus far, there is no evidence of person-to-person spread or of large outbreaks. Small outbreaks may occur when multiple-family dwellings are grossly contaminated with rat and mouse waste.

Various members of the bornavirus genus cause disease in horses, sheep, squirrels, shrews, and psittacine, passeriform, and water birds. The bornaviruses cause severe and oftentimes fatal central nervous system infections. The virus spreads from animals to humans by way of bites and scratches. In the case of a bornavirus infection which came from an imported pet squirrel, secretions in the throat and mouth were loaded with virus. Orf, or contagious ecthyma, is spread mainly to herdsmen from sheep, goats, and reindeer. Hepatitis E spreads to humans from swine, deer, and other herd animals by contact with feces, liver, and raw meat. Newcastle disease virus is spread by domestic and wild birds through occupational exposure.

Novel disease outbreaks, caused by unusual or unknown viruses, almost always come unannounced. Sentinel public health workers are our first defense against the unwelcome interloper. Since everyone in the world now carries a cell phone, even an unofficial sentinel worker and his buddies can call a central authority to get the ball rolling. We can only hope the roads will be navigable and the health centers will have an ample supply of gloves, gowns, masks, IV equipment, and knowledgeable healthcare workers. We have to stem the "brain drain" by training *local* scientists, nurses, doctors, and laboratory technicians and then provide them with employment in their homelands. The damaging effects of climate change on complex microenvironments and enormous animal migrations of every sort are still beyond our control, so for now we must rely on those actions which have been proved to work in the past.

ACKNOWLEDGMENTS

I arrived at Yale School of Medicine in September 1973 to assume responsibilities as a postdoctoral fellow in Pediatric Infectious Diseases. I learned everything I know about infectious diseases, diagnostic virology, and epidemiology from the many learned and enthusiastic people who became part of my professional world—my physician-scientist mentors, my colleagues of both the MD and MD/PhD variety, and our postdocs, residents, and students of both the MD and MPH variety. The University is an extraordinary cradle of intellectual pursuit, and doors all over the campus are open to anyone who wants to learn.

I owe particular thanks to Dr. I. George Miller, who has been Chief of the Section of Pediatric Infectious Diseases during my entire forty-five year career. He has set a very high standard for hard work, clear thinking, pursuit of knowledge, and clinical acumen. I have become an unofficial member of his laboratory family, and he has given me a home in their midst. As long as I've known him, he has focused on the biologic and molecular machinations of the Epstein-Barr virus, the cause of infectious mononucleosis, Burkitt lymphoma, and others. Although I initially worked on a simian model of EBV infection and other smaller projects related to EBV, I ultimately found my calling in the world of HIV and AIDS. Dr. Miller continued to support my efforts in this arena, and he served as a welcome sounding board whenever I wanted to talk about HIV with a "real" virologist and well-trained epidemiologist. Dr. Miller has hosted Thursday afternoon clinical rounds for the past forty-five years. These rounds have become a legend in our Section and in the Department. During these rounds we present the details of our infectious disease consults to one another. We discuss clinical signs and symptoms, social and travel history, laboratory and radiologic test results, and we come up with our differential diagnoses, often outlined on a real blackboard with old-fashioned white chalk. Our discussions are free-wheeling, conducted around a sturdy, eighteen-foot-long oak conference table. We can all see one another in the flesh and listen to one another. To some, this kind of activity may seem sweetly antiquated, but Dr. Miller has taught me and all of us the value of clear thinking, respectful dialogue, and the quest for logical answers in the company of our peers, trainees, and students. No time limits are placed on our conversations.

I especially wish to thank Melissa Grafe, PhD, the John R. Bumstead Librarian for Medical History, for providing me access to some of the older books that rest on the shelves of The Medical Historical Library, one of the finest in the world. The books by Edward Jenner, dated 1798 and 1799, were wonders to behold and to pass my fingers over. Melissa also taught me how to find valuable historical materials online and in some of the various other libraries at Yale.

My publisher and editors have transformed my writings into products that are more readable and more appealing. They've never failed to amaze me with their talents.

I owe a debt of thanks to Yale-New Haven Hospital, its administrators, doctors, nurses, and social workers, for working closely with me as Medical Director of the Pediatric AIDS

Care Program, and allowing me to pursue my dream of creating a piece of the world where mother-to-child transmission of HIV no longer exists. Thanks to their administrative support, their superlative clinical skills, and their love of working together, my dream has been fulfilled. No "AIDS baby" has been born in New Haven for over twenty-one years. I have learned textbooks-full of infectious disease medicine as a result of having cared for infants, children, teens, and adults with HIV/AIDS whose immune systems fail and in whom, as a result, bacterial, viral, and fungal diseases rampage.

I owe Sue Prisley my gratitude for ably turning my scribbled yellow pads of paper into legible prose. It was a job that took hundreds of hours unmarred by even the slightest complaint, not even when some of my stories scared the wits out of her.

Most especially, I have to thank my family for almost all of my professional success and personal contentment. They're an amazing trio of women. Their intellectual fervor, professional excellence, and many other talents make me very happy. I marvel at their many accomplishments and abundant acts of kindness.